SEARCH AND RESCUE

When Disaster Strikes

JOHN MELADY

SCHOLASTIC CANADA LTD.

Scholastic Canada Ltd.
175 Hillmount Road, Markham, Ontario, Canada L6C 1Z7

Scholastic Inc.
555 Broadway, New York, NY 10012, USA

Scholastic Australia Pty Limited
PO Box 579, Gosford, NSW 2250, Australia

Scholastic New Zealand Limited
Private Bag 94407, Greenmount, Auckland, New Zealand

Scholastic Publications Ltd.
Villiers House, Clarendon Avenue, Leamington Spa,
Warwickshire CV32 5PR, UK

Designed by Andrea Casault
Cover image © CORBIS/Graham Wheatley;
The Military Picture Library

Photo credits: page 160

Canadian Cataloguing in Publication Data
Melady, John
Search and rescue : when disaster strikes

ISBN 0-590-51567-5

1. Search and rescue operations — Juvenile literature. I. Title.

TL553.8.M44 1999   j363.34'81   C99-930694-4

5 4 3 2 1   Printed in Canada   9/9 0 1 2 3 4/0

This book is for Bill and Helen Lemaire,
my father- and mother-in-law, with affection and respect

Acknowledgments

I could not have written this book without the help of several individuals who gave of their expertise, time and encouragement. They were all helpful to me. In addition to those whose names appear in the text itself, I would like to thank the following: Harriet and Wally Aksamit, Ted Barris, Mary de Bellefeuille-Percy, Dell Butler, Holly Bridges, Mary and Glenn Butters, Steve Cantrell, Bernie and Theresa Cline, Shirley and Dan Dormer, Danielle Dougall, Leona Hendry and her research staff at the Belleville Public Library, Tim Hanas, Keith Hoey, Chuck Lapp at the National Museum of Naval Aviation in Pensacola, Cathy McDermott, Pete Milnes, Greg Purvis and Commander J.E. Wiggins at the U.S. Coast Guard News Media Headquarters in Washington, Irene and Jack McLaughlin, Bruce Nickson, Tom Nigro, Vince Steiner, Pat Tighe. Special thanks go to Diane Kerner at Scholastic, whose interest in this project was the reason it got off the ground in the first place; to Sandra Bogart Johnston for her editorial skills; to Yasemin Uçar for photo research; and to Andrea Casault for the design. As always, Esther and Dylan Parry were wonderful in displaying their last-minute keyboarding skills. Above all, however, I am grateful to my wife Mary for her advice and encouragement during all phases of the research and writing.

John Melady
Brighton, Ontario
June 1999

Table of Contents

Introduction

Throughout history, humans have looked for the lost, and helped the endangered. If people are missing, we try to find them; if they are injured, we do our best to ease their pain. Of course, if we get lost ourselves, or hurt, we hope someone will find and help us.

But there are people who spend their careers looking for others, and helping those in distress. They are the men and women in search and rescue. They are some of the most highly trained professionals in the world. They have expertise in parachuting, mountaineering, diving and swimming. Their medical training is as highly specialized as that of a paramedic, and they continually update their skills in emergency medical care. They're like athletes who compete in decathlons — exceptionally skilled in a number of different areas. Some search and rescue personnel (often called SAR Techs) work from airports; others from seaports. Some live by the ocean, others rarely see salt water. Some are in the military; others have never worn a uniform. Yet all have one thing in common: they want to save lives.

In Canada and the United States there are many organizations engaged in search and rescue — so many that it is doubtful a single book could include them all. *Search and Rescue: When Disaster Strikes* deals with two of the best known: the Canadian Air Force and the United States Coast Guard.

CHAPTER ONE

Rescue on the High Seas

The terrified sailors were clinging to anything that would save them from being washed overboard . . .

The crowd fell silent as the President of the United States walked into the Rose Garden. He smiled and waved to the invited guests. "Good morning and welcome to the White House," Ronald Reagan began. "I want to make my welcome especially warm, because for so many of you, coming here today was unexpected, to say the least." A ripple of laughter greeted his words. Then he described the sequence of remarkable events that had led up to this moment. It began with a distress call from a Russian ship on the Atlantic Ocean, the *Komsomolets Kirgizii*.

Three days earlier, on the morning of Saturday, May 14, 1987, the United States Coast Guard station at Cape May, New Jersey, had picked up a signal from the ship, which at that time was 340 kilometres off shore. The 147-metre-long freighter, en route from Canada to Cuba with a load of flour, had broken down and was sinking in a storm. The thirty-three men, three women and baby on board would perish if they did not get help right away. Already, two lifeboats had been torn away by the waves. The ship had lost its main power source, its steering capability, and its capacity to withstand the buffeting of the sea. Because Captain Vladimir Khurashev knew his emergency electrical supply would soon give out as well, he ordered his radio operator to call for help.

Within minutes Coast Guard planes from Cape May were airborne. A large four-engined Hercules went first to determine the exact position of the stricken ship, keep the vessel in sight and monitor its situation. But helicopters would be needed to lift people from the boat.

"We were actually out on another search when we were called," said Captain Rick Hardy, a 34-year-old Canadian Air Force exchange pilot flying for the Coast Guard. "The weather was bad and the ship was a long way out. We really didn't think we would have enough fuel to make it out and back, so we were prepared to land on the water if necessary. We sure topped off the tanks before we left though."

About the same time as Hardy and his crew were fuelling their HH-3F Sikorsky Pelican helicopter, two identical machines took off. Coast Guard Lieutenant Keith Comer was at the controls of the first, while Lieutenant-Commander Gary Poll flew the other. Comer was the first to reach the ship, but his flight there was nerve-wracking.

Vicious head winds rocked his chopper, and frequent snow squalls and sleet hampered his progress and visibility. Down below, the cold Atlantic was flecked by freezing spray as the crests of waves fell in on themselves. Foam-flecked black water meant danger was only a heartbeat away. But Comer pressed on.

"I'll never forget our trip out there either," Rick Hardy explains. "My co-pilot, Matt Thomas, was scared. I was scared. All of us were scared. Because the ship was over three hundred kilometres away, we kept watching our fuel. We could fly for about five hours max, and it took us two to get there — without a load. We also knew we would use a lot of gas during recovery. Then, if there were head winds coming back, we figured it was going to be tight."

"It was difficult, demanding," Lieutenant Comer admitted to reporters later. "Being first on the scene, it took us ten to fifteen minutes to determine the best place to hoist from."

Hoisting is search and rescue helicopter crews' most common method of rescuing victims. The helicopter hovers above a site, and a pencil-thick steel cable is payed out by a machine on the chopper. Sometimes a searcher is hooked to the cable and lowered to the person in distress. Other times, a wire rescue basket is sent down instead.

This was the method used to get the people off the Russian ship. Each time the basket came down, one or two or even three people grabbed it, a flight mechanic in the helicopter activated the hoist, and the victim was lifted to safety. In general, the method worked well. But not always.

This time the job was far from easy.

The driving snow and sleet cut visibility considerably. The wind shrieked across the surface of the sea, and when the rescue basket was lowered, it blew sidewise, sometimes on as much as a 45-degree angle. The wind also threw the helicopter all over the sky, and the pilots had to fight the controls in order to remain in one place.

When Lieutenant Comer and his crew arrived on scene, their worst fears about the mission were realized. The ship was nearly awash in heavy seas, listing far to the side, and giant waves were breaking over her deck. The terrified sailors were clustered at the back of the ship — the stern — clinging to anything that would save them from being washed overboard.

The obvious place to pick up the stranded sailors would be the wheelhouse, the highest part of the boat. But the rigging, radar equipment, communications antennae, electrical lines and even the ship's funnel could interfere with hoisting, or snag the hoist cable and basket as soon as it came down. If that happened, the rescue helicopter could become tied to the ship, and be pulled out of the sky as the ship rolled.

Finally, the SAR Techs decided that hoisting would be done from a small, slightly less cluttered deck at the stern.

But then another problem arose. Almost no one on the *Komsomolets Kirgizii* spoke English. Instructions for hoisting were radioed to the ship again and again, but there was no response. Finally the helicopter crew realized what the problem was, and decided to use hand signals to communicate. The flight mechanic stood at the open door of the chopper, pointed to the area where the hoisting would be done, and made elaborate motions with his arm, hoping the Russians would real-

The Komsomolets Kirgizii founders at sea while a search and rescue helicopter struggles to save the people stranded on board. (U.S. Coast Guard)

ize that he wanted them to either get into the rescue basket or at least hold onto it when it came down. It worked.

Comer brought the helicopter in over the pitching freighter. Because his rear vision was restricted, the radio operator directed him into position, and the rescue basket swung out. However, it was immediately taken by the wind, blowing wildly into a huge arc over the ship below.

Comer backed off, gained some height and tried again.

The same thing happened.

At that point someone suggested they drop the basket into the sea in order to stop the swinging motion.

The hoist cable has a reach of about 75 metres, but most of the hoisting was done from about 15 to 25 metres — very low to the water. "All three of us who took people off the ship had to watch our instruments very carefully," Rick Hardy explains. "Because the waves were so high, if you just looked at the sea, the natural tendency was to climb. However, you couldn't climb because you had to be low enough to hoist."

As Keith Comer brought his helicopter down to hoist level, but well to one side of the sinking ship, his flight mechanic dropped the basket into the water. When the radio operator gave the signal, Comer eased the chopper toward the boat. The basket dragged through the waves, bumped up the side of the ship's hull and popped onto the stern deck.

Eager hands grabbed for it.

Several sailors steadied the basket, one of the

women climbed inside, her baby was passed to her, and in no time mother and child were in the air, on their way to safety.

The process was repeated again and again.

"We were in a hurry to get as many off as we could," Comer explained later. "Everyone knew there was no time to spare." In the few minutes since the first helicopter's arrival, the ship had settled noticeably while wild waves continued to slam it.

By the time he and his crew had hauled in 15 victims, Comer knew he could take no more.

"I orbited, waiting," recalled Captain Hardy, in the second helicopter. "As soon as they got fifteen people, they were overweight and had to leave for land. We pulled sixteen people off the boat, and then we were heavy. . . . The plane became much harder to handle, and it wanted to ride nose high, all the way back.

"I suppose we were over the ship for twenty minutes to a half hour. The weather was a mixture of sleet and snow swirling around, and I recall the roughness of the seas, the whitecaps hitting against the side of the ship and washing over it. But we were just too busy to be afraid."

Yet the sailors certainly were.

By the time they were packed into the helicopters, the Russians were terrified. With the shrieking wind, the ever-more-dangerous ocean rollers coming at them, and the roaring of the big helicopters overhead, the situation was unforgettable. The baby who was rescued wailed incessantly, but the adults hoisted into the sky were mostly silent, numb from fear.

"They were pretty quiet," Rick Hardy recalled, "but they did ask if they could smoke. I thought of letting them, but with twenty-one of us packed in there, I said no. There was no argument. They were just happy to be safe. Also," he added with a laugh, "I never really liked the smell of Russian cigarettes."

As soon as Hardy's chopper had cleared away from the sinking ship, Lieutenant Commander Poll prepared to hoist the last six people on board. "It looked as if the vessel was going down at any moment," Gary Poll said later. "The deck was almost totally awash." The last to leave the ship was the captain. Soon afterwards, the *Komsomolets Kirgizii* went down.

The three helicopters flew the rescued Russians to Atlantic City, where they were met by an assortment of government officials, reporters and curious Americans. "There were lots of smiles and hugs," Rick Hardy said. "A few of the sailors could say 'Thank you' and did. They all knew, of course, that they would have died out there. They had not been able to launch lifeboats, and none of them had survival suits. We were also relieved that the mission had been successful. We had no head winds that amounted to much, so our fuel held out on the trip back."

A day or so following the dramatic rescue, while the Russians were being taken on tours of Washington, Coast Guard crews were back at their normal routines: most were busy at Cape May; one or two had time off; others were not even in New Jersey. One of those was Keith Comer. He was in Maine, hauling lumber to an island lighthouse.

Captain Rick Hardy (left) and American Coast Guardsman Keith Comer are congratulated by U.S. President Ronald Reagan. (John Guzman/ U.S. Coast Guard, courtesy of Rick Hardy)

Just before noon on Monday, March 16, he received a radio message telling him he was needed in New Jersey for a press conference. The message and the arrival of a Falcon executive jet to pick him up came as a surprise; yet a further surprise awaited him, as it did the others who had helped the Russians. They were told to be at the White House in Washington the following morning. President Ronald Reagan wanted to give them medals for their memorable life-saving feat.

"Being asked to the White House was a panic," Rick Hardy admitted. "Everything happened so quickly, but it was fun. I think I was just as thrilled as my American crewmen when I got the invitation. I was the only

Canadian, and the only non-Coast-Guard officer on the mission."

Within minutes of their arrival at the most famous residence in the United States, the Russians, the American fliers and the lone Canadian assembled on the Rose Garden lawn. They shuffled nervously in the brief time before Mr. Reagan appeared. But when he did, and when he spoke, they relaxed and listened as he said: "Here we have pilots from Mission Viejo, California; Cincinnati, Ohio; and Prince Edward Island, Canada [who] reached out to sailors from Leningrad, Novgorod and Yaroslavl. I hope and pray that no matter how stormy international affairs are, the leaders of the world can look at what happened between these fliers and sailors and be duly inspired.

"To Captain Khurashev and his crew: Welcome again to our country — and we thank God for your safety. And to all the men and women of the United States Coast Guard who made this rescue possible: As your Commander-in-Chief, it's my high honour to commend you on a job well done." Then the President presented medals to those who had risked their lives to rescue 37 strangers from another land.

CHAPTER TWO

Snatched from a Flood

"The guy's standing in water; we can see his house washing away. . . . Meanwhile, we're running out of gas."

The rain was heavy when the search and rescue helicopter left Squamish, British Columbia. Visibility was limited, and the mountain winds were treacherous. Yet the pilot knew that unless he and his crew went to help, a family of four would be dead by dawn. They were already surrounded by flood water. Their only escape route was closed. And an earthen dam a few kilometres away was on the point of collapse.

It was just after Christmas in 1980. Record high temperatures melted the mountain snow, and the runoff, made even worse by the heavy rain, brought floods. This was particularly true along the Cheakamus River, 80 kilometres north of Vancouver.

The Cheakamus, normally a gentle mountain stream, was now at flood height. In its rampage to the sea it had become extremely dangerous. It threatened several cottages that had been built between it and the nearest highway. In one of these a couple lived with their two children.

Even in summer their home was almost surrounded by water, with their long driveway passing over a rickety bridge. Now the bridge was gone and the people were trapped. A Royal Canadian Mounted Police officer called for a helicopter to lift them to safety. When it

Keith Gathercole piloted the helicopter during the dramatic rescue.

arrived at the Squamish airport, he went with the crew to help locate the home. The helicopter was being flown by Keith Gathercole. With him that night was his co-pilot, John McClelland, along with flight engineers Gerry Gray and Dave McMaster.

"It was really rotten flying weather," Gathercole recalled. "The wind was blowing the helicopter around a lot, and because of the heavy rain and low cloud, it was hard to see. . . . We were also worried about power lines that had been strung across the valley. You could never see them at night." Gathercole decided to follow a highway that he was told would lead to the flooded area and the home they wanted to find.

Gerry Gray described the mood in the helicopter. "The trip was very slow and very cautious," he said. "It was a terrible night with high winds, heavy, heavy rain, and pitch dark. People never get lost or in trouble in nice weather, but when it's really bad, the phone rings."

His friend on board, SAR Tech Craig Seager, also remembered the call for help, and his own drive to the airport that night. "The trip out to the base was one of the worst I'd ever done. The rain was falling in sheets, and a couple of bridges were already underwater."

"John McClelland was reading the map as we went along," Keith Gathercole said as he described the journey into danger. "The floods had taken out the power, so there were no lights anywhere at all. I remember sensing where the bigger hills were; they were the blackest parts outside. We were not sure where the power lines were, so we went pretty slowly.

"We had been flying for a while when we saw a light, what looked like some kind of signal. I briefed the crew as we approached and turned on the lights [the landing lights and the search light] when we got to about three hundred feet [90 metres]. The winds started whipping up again as we got down, and this didn't help much, nor did the trees around this cabin. They were all about a hundred feet [30 metres] high, and were right up close to it. At that time we only had about a hundred and twenty feet [36 metres] of usable cable on the hoist."

SAR Tech Steve Gledhill decided to go down. The military people on board thought that if he couldn't manage a rescue on this wild night, no one could — Gledhill was the best.

As the men in the helicopter lowered him on a steel cable, the rain battered Gledhill and gusts of wind made him swing wildly above the trees. Gerry Gray, paying out the cable, saw with dismay that the swinging was getting worse.

When Gledhill was slightly below the house roof,

the wind slammed him against the building and his feet crashed through an upper window. He ended up sprawled on the wet, slippery roof, clutching at the shingles and trying to get a grip on something. But there was nothing to grab.

He slid down the roof, desperately doing all he could to keep from falling off the roof into the water that surged around the building.

"I felt Steve hit," Gerry Gray explained. "He told us afterwards that there was no grip on the [shingles], so he lost his balance and went zipping down the roof. There were hydro lines leading away from the house and he was sure he'd hit them. But when he ran out of cable, he came to a grinding halt. At that point I thought I'd better get him up out of there."

The second SAR Tech, Craig Seager, peered down at Gledhill, who was still hooked to the hoist. "It was really tough [trying to see] Steve," Seager explained. "We needed our search light, but it was raining so heavily, it was like shining it right on a wall."

Steve Gledhill was hoisted up, but despite his brush with disaster he insisted on going down again. There were still people down there who needed help. He swung back out into the storm.

When he finally got inside the house, he got a surprise: instead of four people, there were two — a man and a woman. It turned out that this was not the family that the police had called about! But Steve Gledhill saved the lives of the stranded pair, then the helicopter flew on.

"Those folks were really stuck," pilot Gathercole

said. "If they had not signalled, we would have never even known about them. They couldn't get out, and in all probability they would have died there. Thank God we saw their light — and all it was was a candle.

"So now, we still haven't found the people we wanted, but the police officer said he knew the river, and if we just followed it along, he would point out the cabin when we came to it. But because of the flooding the river was ten times its normal width; it wasn't in its normal course, and it [seemed] totally different.

"I can remember him looking down at what looked like a lake and saying 'I don't know where I am.' The water had risen so much during the night. But around about that time Craig Seager thought he saw something at a river bend. We had actually passed over the spot, but when he yelled, we turned around and went back. It turned out that this was the family we'd come for in the first place."

This time Craig Seager went down. The cabin was surrounded by towering trees swaying in the wind, and the cable wasn't long enough to get him to it in a direct line. Instead, he decided to wade in, and had Gray drop him into the raging river in front of the cabin.

"I had no idea what happened," said Gerry Gray. "It was so dark, and the search lights weren't much help. I was waiting for [Craig] to signal that he was down, but he never did, and then the cable got tight. I figured something was wrong, so I pulled him up again."

"I couldn't get close enough to the house the first time down," Seager added. "As soon as I got into the river, the current was too strong to fight against. I remember hitting the water and I was gone like a flash. So they brought me up and decided they would have to hover down wind, which they don't like to do because they have less control that way. But it was the only way they could get me close to the house.

"When I got back down again, I got behind the house. The house was upstream from me, so there was some protection from all the junk coming down in the current. I was wearing a wetsuit of course, and the water was chest deep. I worked my way up to the house, and I could see the people in a window above me. I told them there was a dam upstream that was in danger of breaking, and that we were there to get them out."

Seager took up the mother and baby in his first-ever triple hoist. Then he went back for the three-year-old. But the boy refused to come.

Seager and the boy's father did everything they could to persuade the child, but nothing worked. He was absolutely not going to go with this strange man, up into the sky, in a storm, in the middle of the night. In exasperation Seager grabbed the boy and put the rescue collar around him. But the adult-sized collar just slipped off over the small child's head.

Finally Seager just grabbed the terrified boy with one arm, held the cable with the other and signalled for a lift.

The first few metres were okay. Then the boy started to scream and flail around, pounding and kicking with all his might. Seager held the boy closer to ward off the fists, but the kicking was not as easy to prevent — or ignore. The trip up seemed to take forever. Two

Craig Seager (right) who played such a vital role in rescuing the stranded family, has won both the Medal of Bravery and the Star of Courage. (Canadian Forces photo)

metres from the safety of the chopper door, the boy's struggling increased. Had it not been for the strong arms of Flight Engineer Dave McMaster, waiting to pull the two in, the boy might have broken free and fallen to his death. McMaster deposited the hysterical child in his mother's arms.

Now there was just one more rescue to go. But nothing was going to go smoothly this night. With Gledhill on the end of it this time, the cable began to swing out of control in the increasing winds. Gray, in the chopper, could not stop it, nor could Gledhill.

Gerry Gray's help from inside the helicopter was key to the successful rescue during the 1980 Cheakamus floods.

Keith Gathercole reversed the helicopter, but he had almost no room to manoeuvre in the tight space of the clearing. With one wild swing, Gledhill crashed into the top of a pine tree and spun around it. The cable wound tightly around the tree, and the SAR Tech was trapped, clinging to the tree with the rope wound around the trunk above him.

Now the helicopter was tied to the tree, and all its personnel and passengers were in danger too. Luckily, Gathercole was able to keep the machine steady until the crew worked their way out of the dilemma.

"Gerry Gray was the consummate professional in the whole thing. He was so calm," Gathercole recalled. "Because I couldn't see down to Steve, Gerry directed me back and forth, sidewise and back until we actually unwound the cable. He'd say, 'Okay, move it back a yard [about a metre], now right a yard, now ahead two yards,' and so on. I had a big pine tree brushing the window for reference, and the unwinding seemed to take a long time. All the while, poor [Steve] is still down there holding onto the tree."

Finally Gledhill managed to free himself and Gray lowered him to the ground, and within minutes he signalled to be brought up. But he came alone — the man refused to come.

"We couldn't believe it," Gathercole explained. "The guy's standing in water; we can see his house washing away, and there are big fir trees, roots and all, floating downstream. As we watched, parts of the house fell down. Meanwhile, we're running out of gas."

They all knew the man would be dead in minutes unless he was pulled from the river. Gledhill turned to the man's wife for advice. "Just go down and make him come!" she ordered.

Gledhill grabbed the hoist as the fuel warning lights flashed on in the cockpit.

This time Gledhill gave the man no choice: he just wrapped the collar around him and signalled Gray to lift. In the meantime Gathercole and co-pilot John McClelland prepared to make an emergency landing.

"All this time, I'm looking around trying to figure out where I could put down," Gathercole explained. "I thought we'd never get back to town; there just wasn't enough gas. But John was calculating our fuel situation, looking at a map, and trying to decide if we could make it. Then I looked at him, he looked at me and we made the decision to try.

"Luckily, it was almost dawn, so we could more or less see where we were going. Thank God, we made it — just."

The long night was over. The crew got some rest,

their helicopter was refuelled and they took off to search for more victims of the Christmas floods.

During the floods a long stretch of CN railway track was left suspended above a washed-out section in the Fraser Canyon. (Bill Keay, The Vancouver Sun)

CHAPTER THREE

Slammed into the Mountain

"I've got it. I've got the crash site!"

Pilot Mike Ruwald waited for takeoff clearance from the Edmonton control tower. With Ruwald that day, April 1, 1983, were his wife, Bonnie Boucher, and their friends, Deborah and Luke Reiche.

The group was headed to Vancouver in a four-seater Piper Comanche which Ruwald had rented in Edmon-ton that afternoon. It was a small but comfortable single-engined plane and, luckily for its passengers on that day, was equipped with an Emergency Locator Transmitter.

Ruwald's flight plan took them west-southwest over the still-frozen prairie. Off to the right the passengers could see the Yellowhead Highway running west toward the Rockies, and to the left the road to Calgary stretching to the south.

In no time the Comanche was climbing above the foothills of the Rockies, then high enough to clear the mountains themselves. On a clear day the scenery here is spectacular, but the same elements that make it magnificent also make it treacherous to fly over, particularly in a small plane that cannot climb too high. Most of the time Ruwald flew in cloud, using IFR, Instrument Flight Rules. But he was an experienced pilot and his passengers trusted him. They knew they would soon be over the mountains, out of the clouds and touching down in Vancouver.

They were wrong.

All went well on the first leg at the higher elevation. The altimeter indicated adequate clearance over the highest peaks, and the clouds that enveloped the plane made it seem like flying in a white cocoon. The flight was smooth and they made good progress.

Soon they were beyond the Great Divide, past the Monashee Mountains and above the Cascade Range that runs north-south, paralleling the Trans-Canada Highway between Hope and Lytton, British Columbia. Suddenly Ruwald sensed a slight hesitation in the engine. He checked the gauges but everything seemed fine, and soon the problem seemed to vanish.

But then the hesitation returned, and this time everyone noticed it.

Ruwald did more checks and reassured his passengers. He also flipped the carburetor heat switch on because he thought the trouble might be an ice build-up there. This helped for a moment. But again the hesitation returned, and the engine seemed to be losing power.

Now all four people in the plane were worrying. Ruwald called the airport at Abbotsford, ahead and to the south of his position. He described the problems he was having and told the tower that he was trying to correct them. He also reported his position. As he finished his transmission he saw that in losing power he had also lost height. The situation was getting worse.

Now they were in the Lillooet Range of mountains, and Ruwald could not afford to fly lower. Yet the problem continued. The engine seemed to almost quit, roared back into life and then quickly sputtered as badly as before. At that point Ruwald called Mayday. Shortly thereafter the engine stopped. The Comanche slammed into the side of Chehalis Mountain at 1600 metres.

✧ ✧ ✧

"I was a SAR Tech at Comox at the time," said Arnie Macauley. "We got word from RCC [Rescue Co-ordination Centre] that there was an aircraft down and that they had picked up an ELT [Emergency Locator Transmitter] hit. Later, one of our search planes pinpointed the exact position of the crash, on the east side of a mountain, sixty miles [100 kilometres] or so from Vancouver. We learned later that the pilot had lost his engine in cloud because of ice, and didn't make it through the mountains. The crash occurred late in the afternoon, so there was no time to get in to the site before dark. We launched in a Lab at first light the next morning."

With Macauley in the Labrador helicopter were the pilot, Major Reg Lanthier, and George Makowski, another SAR Tech. The weather was so bad it took most of the day for the crew to reach the base of the valley near the crash site. "A couple of media helicopters came in with us," Macauley said, but, "it was still a while before we got a small break in the clouds. As soon as it came, we fired up right away."

Lanthier lifted the big Labrador into the sky. As they approached the crash site, the ping of the ELT grew louder. The men inside the rescue chopper were mostly looking for wreckage at this point, because no one held out much hope of finding survivors 24 hours after the plane had gone down. Even if some had lived through the crash they would probably have perished from exposure. A ground search had to be called off because of the rugged terrain and heavy snow.

"We were still climbing," Macauley continued, "when we flew by something that stood out. I remember our flight engineer, Dave Davis, spotting the wreckage and saying, 'Got 'em.' Then George, who was sitting at the right window, yelled, 'I see three survivors!' We told him he was crazy, but he said, 'No. There's three people down there waving out of a snow cave!' We could hardly believe it. They were on a sixty-degree snow slope, and it looked as if the airplane had crashed right into it. Then it had flipped upside down. It was obvious that there had been snow overnight and it looked as if there could be an avalanche at any minute. George and I wondered how we were ever going to get them out of there."

Everyone on board the helicopter that day knew that if the people on the mountain were not picked up, they wouldn't make it till morning. The very fact that they had survived to this point was incredible. They had endured the crash, a major snowfall, freezing temperatures and the imminent risk of avalanche. Making it through another night would be a miracle — and the weather was closing in again.

"We first decided we'd get in as close as we could to the snow slope and hoist from there," explained

Arnie Macauley beside a Labrador search and rescue helicopter. (Canadian Forces photo, courtesy of Arnie Macauley)

Macauley. "So we pulled in to about sixty yards [over 50 metres] from the mountain, but just got blasted out of there by downdraft. We peeled off and changed our plan. This time we decided to put on our mountain gear, have the helicopter put us down on the top of the mountain, and . . . climb down to the crash from there. But the aircraft was right at its limit and there was so much turbulence up there it was hopeless. We had to pull out. Then we wondered about going down to the valley, dumping all of our extra equipment and trying with a lighter load. But then we realized with the size of the Lab there would be so much vibration and prop wash we could blow them right off the mountain and start an avalanche as well.

"Finally, we figured that if we had a helicopter that could hover at the altitude of the crash and create less turbulence, we might be able to get the people. So we called RCC and told them to find another helicopter right away. While they were doing that, we went down to the valley."

A CBC camera crew had their helicopter sitting there when Macauley and his crew landed. "One guy came over and asked if they could be of any help," Macauley said. "I figured I could use [their aircraft] to have a better look at the crash site, to try to work out our next move. So the pilot kicked the cameraman out and I hopped in. We hovered about thirty metres in front of the crash, and I got a better look at it. I was also able to identify some features on the mountainside. As it turned out we were lucky to get this second look because the weather was coming down and clouds were already back over the site."

Macauley and the pilot of the media helicopter then returned to the makeshift base camp in the valley. They had barely arrived when another chopper came into view. This was the replacement aircraft the RCC in Victoria had found on such short notice. The machine was owned by a British Columbia company, Okanagan Helicopters, and was being flown by its chief test pilot, Terry Dixon. The craft was a Long Ranger, a more powerful version of the better known Jet Ranger, and the man flying it was said to be a top-notch pilot.

"Terry showed up with the Long Ranger and a couple of slinging cables," Macauley said. "Fortunately, he had stopped in Abbotsford and picked up Craig Seager, another of our SAR Techs."

Seager was both a veteran and a highly skilled rescuer. He was often said to be fearless, and already wore the ribbon of the Star of Courage, Canada's second-highest award for bravery, on his uniform. Just a day earlier he had plucked a seriously ill man from the bobbing deck of a sailboat.

"We sure needed him," Macauley said. "We jury-rigged a Billy Pugh net [a large mesh scoop/cage rescue device] on a cable underneath the helicopter, and Craig Seager and George Makowski got in it. I rode in the cockpit to direct Terry.

"We took off with the Billy Pugh hanging below us, and Craig and George swung out three thousand feet [900 metres] or so over the valley. By this time the mountain was covered in cloud, there were ice crystals blowing around, and we could hardly see a thing. But we had all the luck in the world. I found this ridge of rock that I knew led up to where the people were, so we followed it and found them."

When Macauley located the ridge the helicopter was about 100 metres below the crash site. Dixon had to manoeuvre extremely close to the cliff face, with the tips of the rotor blades whirling frighteningly close to the mountainside. One mistake and a blade could touch and send the helicopter spinning out of control, killing the rescuers and leaving the crash survivors still stranded. That afternoon Terry Dixon lived up to his reputation. He kept his cool and flew magnificently.

Once during the slow, tense climb, Dixon turned to

Macauley and said: "I'm getting nervous. Did you feel that?"

"Feel what?" the SAR Tech asked.

"My heart. When I'm nervous I can feel my heart pounding."

But Macauley said the chopper was steady as a rock. Dixon never lost his focus. He flew the Ranger, kept an eye on the rotors and slowly climbed the cliff face. Macauley did his best to look out for Seager and Makowski in the net below the helicopter, while at the same time watching how close the tail rotor was from the cliff face so he could direct Dixon out if the tail end was getting too close. There was another ridge below and behind them, and no one wanted to hit it if they had to back off suddenly.

"Then Terry called, 'I've got it. I've got the crash site,'" Arnie Macauley said. "He tucked the [chopper] in even closer, it seemed, and Craig and George were able to swing the net back and forth and jump onto the snow. They had their ice axes with them, so they got a good grip. If they hadn't, they would have fallen all the way to the valley floor."

When Seager and Makowski reached the wreckage of the Comanche they found Bonnie Boucher and the Reiches alive and — considering their harrowing ordeal — in relatively good condition. Luke Reiche had broken an arm in the crash. Pilot Mike Ruwald was dead.

✧ ✧ ✧

Quickly, the SAR Techs strapped the two women into the rescue cage, Makowski held onto the side of it, and

Dixon prepared to take the first load down. Seager and Luke Reiche remained at the crash site.

"Just getting out of there was dicey," Arnie Macauley said. "The cloud was like soup and we could hardly see a thing. Terry felt he couldn't just back down the mountain, but we had to back away from it. Then he did a pedal [left rudder] turn . . . while I tried to watch the ridge on our right. Finally we got out of there, popped through the cloud into clear air and whisked right down to the valley."

"While they were gone, only twenty minutes or so, I kept a pretty close eye on Mr. Reiche," said Craig Seager. "I checked him over, splinted his arm and kept asking him how he was doing. I knew he was cold, and I'm sure he would not have lasted another night up there. I remember telling him that they would be back for us, but even if they could not get back, I was going to stay with him in any event. I told him he would be okay."

Back at base camp, Makowski and Macauley put the two women into electric blankets in the waiting Labrador, then returned for Reiche and Seager. Fortunately, by this time the cloud had cleared.

Because the priority in search and rescue is to assist the living, the body of Mike Ruwald had to be left on the mountain when Luke Reiche and Seager came down in the Billy Pugh. The three crash survivors were immediately taken to hospital in the Labrador.

"We really admired those folks," Macauley said. "They had some cuts and scrapes, and the broken arm, but after what they had been through, they were in

Arnie Macauley, George Makowski and Craig Seager (left to right) receive the Medal of Bravery for their courageous rescue. (Tim Smith/Canadian Forces photo)

pretty good shape. They had done everything right. After the crash they dug themselves the snow cave, then hauled seat cushions out of the airplane. They poured some oil over the cushions and kept a fire going for warmth. By doing all this they had saved their own lives. But they were about at their wits' end when we got there. The oil was gone and they were already getting hypothermia, and were starting to lose it."

The survivors were treated and released from the Abbotsford hospital, and by chance ended up staying in the same hotel where the SAR Techs had rooms.

"Later that evening I heard a tap on my door, and when I opened it Bonnie Boucher was standing there," explains Macauley. "She told me the RCMP did not have the equipment to get her husband's body off the mountain until the weather was better, and she asked if we would bring him down for her. So George and I volunteered.

"The weather was good the next day, and this time Terry brought another pilot with him. George and I took some rescue tools and crash axes with us, and even then it took us over an hour to get the body out. Because the plane had flipped when it crashed, it came down on Mr. Ruwald and had broken his neck. We actually had to cut through the floor to get to him. When we had the body ready to go, we collected it, the logbooks, the luggage and anything else of value, and put everything into a big cargo net. Terry took this down first. Then he came back up with the Billy Pugh and George and I had a nice ride down."

Three years later Arnie Macauley, George Makowski and Craig Seager stood in the ornate ballroom at Rideau Hall in Ottawa as the Right Honourable Jeanne Sauvé, Governor General of Canada, presented Terry Dixon with the Star of Courage for an act of "conspicuous courage in circumstances of great peril." A few minutes later the Governor General turned to Macauley, Makowski and Seager and bestowed on them the Medal of Bravery.

CHAPTER FOUR

Channel of Terror

*"We could hear the hissing sound of the flames as they
came toward our boat. . . . Then a huge wall
of flames went ripping by us."*

When several days of torrential rain hit east Texas in
mid-October, 1994, nearly 20 people were killed.
Millions of dollars worth of property was destroyed.
Paul Angelillo remembers it well. He was there.

"There" is the Houston Ship Channel, a waterway
that runs from the city itself to the Gulf of Mexico, 90
kilometres away. Ordinarily it is over 60 metres wide,
and deep enough to handle the volume of ocean-going
vessels that move through every day. It is busy, vital
and almost never closed.

"I was with the Coast Guard at Galveston at the
time," Angelillo explained. "We were responsible for
the ship channel, and it was our job to keep it open.
That involved water patrol, looking out for accidents,
placement and repair of navigational aids — that sort
of thing. I found the work hard and often pretty dangerous. It was basically a very, very nasty place to work
— but I loved it! . . . You had to push your own limits,
but it was fun, and I think I contributed. I also learned
how to really drive a boat. That sure helped during the
floods in the fall of 1994."

The fall in question had been unusually humid in
east Texas, and it was unseasonably hot well into

October. There had been vicious thunderstorms earlier on, but the rain was fleeting. The downpours were gone as quickly as they came.

But then there was Rosa.

Rosa was a tropical storm that had formed off the Pacific coast. Moist winds from her swept across the southern states, northern Mexico and the Gulf. But they stalled when they came to Houston because a cold front from the north collided with the warm southern cloud mass and the weather systems mixed. Together they produced rain — lots of rain — all at once. Most of it fell on Houston and the counties surrounding it.

"It kept raining and raining and raining," Angelillo said, "and it wasn't a gentle, put-you-to-sleep kind of rain; it was heavy, heavy, and with it, thunderstorm after thunderstorm. . . . Pretty soon everybody began to get a bit panicky, and the first flood watches came out. That was when everything started to go haywire. We were told to take our boat and go up north, up the ship channel, all the way to San Jacinto to see what was happening. That would have been almost forty miles [65 kilometres] or so from where we were. We were told to monitor what was going on, and to check on the currents."

At about the same time as Angelillo and his crew were preparing to leave, officials to the north of Houston, near Lake Livingston and along the banks of the Trinity River, were preparing for the worst. Water levels had risen alarmingly, and some low-lying areas were already flooded. And still it rained.

Evacuation orders were issued, at first to hundreds of people; within hours, to thousands. Many did not wait to be told to go. Others refused to leave and had to be carried from their homes. The Red Cross opened evacuation centres, eventually more than 50 in all, and in some cases even these had to be abandoned as the water rose to levels higher than ever before.

Flash floods became commonplace. Dry ground would be underwater in an hour. People became stranded, stuck and separated from their families. Some died. Two men who attempted to drive through a low spot on a rural road ended up in a lake. Hours later, only one was found. A mother and father had their minivan washed from a bridge. Both survived, but the baby the mother held was drowned. A man's pickup was lifted by the torrent of a flooded creek. It flipped over and the man was dead when he surfaced. A woman and the four children in her car were swept from a highway. Two of the children died. Helicopters and horses had to be used to retrieve their bodies.

Near La Porte, just east of Houston, 14-year-old Larry Joe Lackey decided to go for a swim in the deep, warm waters that swirled down his street. But the current engulfed him, and he died when he was sucked into a storm drain. Not far from a dam at Lake Livingston, a man on horseback, trying to rescue a woman from her house, perished when his horse stumbled in deep water and he was torn from its back.

Soon the situation in Houston began to attract attention elsewhere. Governor Ann Richards flew over the floods in an army helicopter and declared 48 counties eligible for state disaster relief. President Bill Clinton offered help from Washington. Radio, television and newspaper reporters descended on the area.

But still the waters rose, and as they did, other problems became apparent. Storm drains were so choked that the flow was reversed. In some cases rats that lived in these sewers were washed into homes, where they emerged from toilet bowls. Hundreds of snakes slithered everywhere. One man reported running over a whole mass of two-metre-long water moccasins with his truck.

"It was really a miserable situation everywhere," said Paul Angelillo. "The first thing we noticed when we got going in our boat was the current. It was phenomenal, and there was all kinds of debris and junk coming down. The water in the channel is always polluted, but now it was really dark, like dirty chocolate milk. You couldn't see anything at all under the surface, and that day and the next, we hit so many things — the tops of stop signs, car roofs and so on — that our propeller had to be replaced several times.

"There were three guys who worked with me: Dustin Wagner, Brad Venendaal and Matthew Knight, and we were all busy, all the time. We went pretty well flat out for forty-eight hours straight. At the end of it, we were really beat."

By the time Angelillo and his crew got their seven-metre aluminum Coast Guard boat into the channel and then as far as the mouth of the San Jacinto River, directly east of Houston, Interstate 10 was barely above water. "Ordinarily there would have been ten yards [10 metres] or so clearance from the river surface to the bridge there, but now there was about half that. On either side of the bridge were bulwarks, where the water would normally flow through, but they were blocked with all the debris. The water was all funnelling through the centre, and it looked like a giant faucet that had been turned on. We had to literally go uphill to get through."

Angelillo brought his boat out into the middle of the river, told his crew to secure everything and hold on, then opened the throttle wide. The big 270-horsepower engine erupted in a roar, its turbo-charger kicked in and the boat plowed up the face of the cascading waterfall. "As we shot under that bridge, I just prayed to God that we wouldn't have a breakdown," Angelillo re-called.

By this time the four men, along with dozens of others, had received instructions by radio to start evacuating people from the worst-hit flood areas. One of these was a housing development on a bend in the San Jacinto River. "The river had been diverted and all these homes had been built there," Angelillo explained, "but with all the rain, the river had just carved out a whole new channel, which had actually been its original path. But some folks who lived there were pretty stubborn. They told us they had survived floods before and they would ride this one out too, but it was obvious their homes were going to go, so we had to get them out. Many went without an argument when they realized what they were up against, but many didn't. Anyway, we did our best."

Several homeowners had to be rescued from porches and balconies, and as the water rose ever higher, from rooftops as well. "And they brought all kinds of things with them," Angelillo said, "dogs, cats, parrots, hamsters — every pet you can imagine. Some had suitcases, some had nothing.

"A lot were pretty upset. That included moms and dads, which made it particularly hard for their kids. Ordinarily children don't expect to see their parents afraid, almost to the point of tears, but they did that day. Often we were uneasy about the situations we were in too, but we tried to keep it light, tried to assure them that everything would be all right. The last thing they needed was to see scared rescuers."

One of the people who had to be rescued raised emus, the big flightless birds of Australia. As the San Jacinto waters rose, several emus broke loose and raced along the riverbank in a panic. A Texas park warden attempted to corral them, intending to get them to higher ground, but had to wrestle with a particularly stubborn one.

"The emu didn't want to co-operate at all, and it kicked the guy, right in the chest, and sent him flying," Angelillo said. "Paramedics had to come in and get him out. We decided to ease off on the emus."

Other rescues were traumatic.

"Late that first night we were diverted to a house where an old couple lived, at a place called Banana Bend, on what had been a rather tight loop in the river. The current was unbelievable," Angelillo recalled. "It was roaring in our ears, and we had trouble getting in beside their house. The whole place was shaking, and when we put our spotlight on it in the darkness, you could see it tilting to one side. It was about to crumple.

"Now, you always work with the current in front of you, to give you more steerage. I did and then let off on the throttle a bit, but we kept a spotlight on the house and on the man and woman we were trying to reach. Finally we got to them, got them into our boat . . . and their house disappeared before our eyes, in thirty seconds!"

Another evacuation was unforgettable for several reasons. Shortly before dawn, in an area that was completely flooded, someone noticed a flare in the sky above a house that had not yet been checked. The house was built on stilts, and these were swaying from side to side as the muddy torrent clawed at their base. It was obvious the whole place would soon collapse, and take anyone inside with it.

"We found one man living in the place," Angelillo said, "and he had fired the flare. But getting him out of there was something I'll never forget. It was one of the stickiest parts of the whole night. Actually, we even had trouble getting to the guy's house. There was really no way to navigate straight in. We had to use a line to tie off the boat, then play it down the current to the place. That's when our fun began."

Once they had secured their boat, Angelillo and Matthew Knight climbed up rickety stairs to the front door. Brad Venendaal, still in the boat, shone a searchlight so they could see their way. The men pounded on the door.

No response.

They pounded again. Still no response.

They knocked a third time, and got the shock of their

Paul Angelillo kept his cool even when a madman wielding two revolvers refused to be rescued.

lives. A grizzled, obviously drunk, wild-eyed man ripped open the door and roared, "What do you want?"

Angelillo was about to answer when he realized the man had a gun in each hand, and both weapons were pointed at Angelillo's chest.

"To say I did a double-take would be an understatement," Angelillo said. "Matthew and I backed off, but we didn't back down. Neither of us was armed, and because we were completely at the mercy of this guy, I figured I had better start explaining who we were and why we were there."

He lowered his voice, willed himself to remain both calm and outwardly nonchalant, then told the homeowner why they were there. He also suggested that perhaps the man could lower the revolvers he held.

The man did not react.

Angelillo tried again. He told the man how bad the

flooding was, that they were concerned for his safety, that they feared his house would be washed away. They had come to help him in any way they could — to rescue him. They invited him to come in the boat and they would take him to a shelter where he would be safe.

The response was a torrent of swearing and the continued pointing of the guns.

Paul Angelillo paused, then suggested that the man lower his guns and let Matthew and himself come in so they could talk. "I was playing for time, trying to think of what to do next. Obviously, neither Matt nor I wanted this guy to shoot us, but we sure didn't want to be at the house much longer either, because it was about to collapse."

Finally the gunman lowered his weapons and invited the Coast Guardsmen inside. Then, for almost two hours, the three talked about the situation at hand. All the while, the house was shaking, the men in the boat outside were wondering what in the world was going on, and the raging waters of the San Jacinto were rising.

"The first thing we noticed when we got in there was that there were weapons all over the place," Angelillo recalled. "We saw shotguns, rifles, lots of pistols, a grenade launcher — even a flame thrower! And this guy was acting like a crazy man."

Still, negotiations continued. While Knight took his turn talking, Angelillo assessed the man, trying to decide if he could be overpowered — and if need be — dragged out the door and taken away. But then, rather unexpectedly, he agreed to go. At first he wanted to take some of his guns with him, but when his rescuers said he could not, he agreed to leave them behind. He was removed from his house just as it collapsed. Minutes later the building and all the guns that had been inside washed away toward the Gulf of Mexico. The man was quickly handed over to the police. His rescuers went on to save others.

Their boat headed for a fire that had broken out when two gasoline pipelines were ruptured. There were several such pipes in the area. In most cases these had been buried under at least a metre of soil, and they were probably as secure as it was possible for them to be — in ordinary circumstances. But during the floods, nothing was ordinary.

One of these huge pipes, more than a metre in diameter, and owned by the Colonial Oil Company of Atlanta, Georgia, came apart because of its own weight when the soil under it washed away. Over a million barrels of gasoline a day passed through the pipe. When a spark of unknown origin ignited the gasoline, the resulting firestorm was awesome. A house, a railway bridge and several boats were burned by flames as high as a ten-storey building. Smoke was seen 50 kilometres away. Several people were hospitalized with burns. The Houston Ship Channel was closed.

"We could hear the hissing sound of the flames as they came toward our boat," Paul Angelillo recalled, a shiver in his voice, "and that was over the sound of our engine. Then a huge wall of flames went ripping by us, not twenty feet [6 metres] away. At that point, we realized we were of no more help to anyone. There was nothing more we could do, so we got out of there fast."

The Coast Guard boat was beached, and the men who rode in it sprinted to safety.

The fire was allowed to burn itself out . . . and in due course the flooding ended.

*Gasoline and crude oil from burst pipelines ignited, sending flames and smoke billowing into the sky. Several buildings were destroyed when they were hit by burning patches of gas and oil. (Kerwin Plevka/*The Houston Chronicle*)*

CHAPTER FIVE

A Deadly Stall

Suddenly the plane shot out of the clouds into a clear space.
Right in front of their eyes, not a hundred metres away,
was the tree-covered side of Mont Jacques Cartier.

The weather warning from the airport manager did not dampen the spirits of the three young people heading home from their camping trip in a rented Cessna 172. The sun was shining, the breeze was soft and the weather was pleasant. The pilot, 18-year-old college student Christian Godin, knew he had to keep clear of the highest point in the area, the cloud-covered Mont Jacques Cartier. He had filed a flight plan that would take them well to the south of it.

Christian had been flying for a year and a half, and hoped someday to become a commercial pilot. He loved flying. With him that day were his sister Edith and his good friend Monique Seguin. The date was July 24, 1977. Takeoff time was 1:00 p.m.

Godin planned to follow a fairly straight line westward, over the rivers and rocks of the Gaspé, the beautiful forested peninsula of eastern Quebec. His first stop would be Mont Joli, a small town on the banks of the St. Lawrence River. There the plane would be refuelled; then Godin and the others would set out for Montreal and home.

For some reason, just before leaving the terminal, Christian bought some insurance so that he, his passengers and the plane would be covered, in the unlikely event of an accident. The insurance, he hoped, would be a guarantee of a safe passage home. The purchase was made on the spur of the moment, but as he would admit later, it was one of the best impulse buys he ever made.

The first part of the flight went well. Except for a few air current bumps after takeoff, the plane's progress was smooth. The steady drone of the engine was reassuring. But the area below the plane was largely wilderness. Apart from a lonely two-lane secondary road that led for a while in the general direction he was headed, Godin had to trust his instruments and his experience to get where he was going.

Eighty kilometres from the departure point, the plane entered the air space above a provincial park. The preserve was large, undeveloped and remote. It was the land of lynx, beaver and bear. Few people knew it well. Those who did were aware that its dominant feature was the 1248-metre-high Mont Jacques Cartier.

For some reason Godin misjudged his direction in the park. The Cessna began to drift to one side, about 20 kilometres from the track indicated in his flight plan. And even though he did not notice that he was off course, Godin did notice a cloud bank up ahead.

He had flown in cloud before, but close to home — where he could locate familiar landmarks, accessible airstrips and nearby traffic control centres where he could talk by radio with people who could give direction. Here was just cloud, more cloud, and then the first few drops of rain . . . and the hidden peak of the highest mountain around.

Christian, Edith and Monique became more and more worried. They looked everywhere, but all they could see from the windows was a grey-white nothingness. It was as if they were inside a great ball of cotton batting: no colour, no contrast, no clarity anywhere.

Christian constantly checked his map, air speed and altimeter (the instrument that told him how high they were) but they did not give warning of anything ahead. In fact, there came a point when he did not know what was ahead.

"The clouds boxed us in," he said later. "I started climbing because I knew there was a mountain around, but I didn't know where."

Suddenly the plane shot out of the clouds into a clear space. Right in front of their eyes, not a hundred metres away, was the tree-covered side of Mont Jacques Cartier!

Christian yanked the control column of the Cessna back. The nose of the plane came up, but the abruptness of the movement caused the motor to stall. In the words of so many veteran pilots, this was the point where Godin ran out of altitude, airspeed and luck. All at once.

The plane didn't actually dive into the trees; it more or less settled into them. Branches brushed the bottom, the wings and the sides of the fuselage. The propeller, still spinning, dug into pine boughs, twisted out of shape and stopped. A tree limb crashed through the windscreen, but miraculously missed the people in front. Windows on either side disintegrated. Then a huge rock under the right passenger seat tore through the cockpit floor and smashed Monique's right leg in four places.

For a few seconds no one moved. In the sudden silence the three began to realize what had just happened. The Cessna settled to the ground, remained right side up, and lay there like a broken bird. As quickly as he could Christian unbuckled his seat belt, then turned to assist his passengers. Monique, it was obvious, was badly hurt. Edith seemed better. She was able to help her brother drag their friend out of her seat and away from the wreck. All three feared fire, but it never came.

The adjustment to the sudden turn of events took time. Monique was half-carried to a fairly level spot where she would be as comfortable as possible. Christian's head was cut, but only slightly. Edith complained of sore ribs. All were in various stages of shock.

They stayed close to the plane, but at the same time tried to figure out where they were. "There was no panic," Christian told others later. "There was no time to react. It all happened so quickly, like a car accident. The radio was knocked out and the windows were all smashed. We crawled out and set up our tent."

Then they settled in to wait for help.

A couple of hours later, the people on duty at the airport at Mont Joli began to realize that the Cessna 172 they were expecting had not arrived. Checks revealed cloud and rain to the east, so it was assumed that Christian Godin might have tried to fly around troublesome weather. But as time passed and still no

Cessna arrived, concern grew. Finally officials at the Rescue Co-ordination Centre at Halifax, Nova Scotia were alerted. They sent a plane to have a look.

"I was based at Summerside, Prince Edward Island at the time," SAR Tech Rick Quesnel explained. "Late that Sunday we were told that a plane was overdue, and that it might have gone down. We were also informed that it did not have an ELT so we would not be able to pick up any radio signals from it.

"We flew up to the Gaspé to have a look, searched for [almost two] hours but found nothing. . . . The next day we searched again, but again nothing turned up."

The second day an Argus, an anti-submarine patrol aircraft, was brought in along with helicopters and a couple of Buffalo airplanes. On board each were several spotters whose job it was to observe the terrain below and try to see if they could locate a downed plane. The job of spotting was extremely tiring because the landscape looked the same wherever the planes flew. There was rock, forest, lakes and little more. Spotters looked not only for the plane, but for wreckage, smoke, broken trees, signal fires — anything that might lead them to the location. They worked for only 20 minutes or so at a time, then rested while others took over. The constant sweeping of one's eyes over the endless forest is almost hypnotic, and when spotters are exhausted they can easily miss what they desperately want to find.

By the time dusk had settled on that second day, no plane was found. The crash victims were still up on the mountain.

"Because the people who were missing were really only kids, we were under a lot of pressure to find them right away," recalled SAR Tech Mike Johnston, who was searching from a Buffalo. "While we were airborne the third day, a message came over the radio saying that a crash had been sighted. Somebody in the Argus saw it and they called us in."

Searchers in the Argus had found Godin's plane lying wrecked on the slope of Mont Jacques Cartier, not far from the top. No one knew the status of the survivors, or if there even were any. The sooner somebody got down to the plane, the better, and the nearest jumpers were on the Buffalo.

"The terrain near the top of the mountain was really steep," said Mike Johnston. "There were sharp ravines and valleys on every side, and the whole thing was covered with a lot of cloud."

"And the winds were in excess of thirty knots [55 kilometres per hour]," added Johnston's jump partner, SAR Tech Paul Beattie. "We knew that getting down there might be tricky."

Pilot Nick Rapagno passed back and forth over the mountaintop many times. Because of the high wind, the cloud and what looked like the smallest of clearings in the underbrush, everyone on board knew that any descent would be dangerous, and would require both skill and timing on the part of those involved. SAR Tech Rick Quesnel, who did not jump to the scene but who came in by helicopter later on, said that the jumpers could easily have been killed on the mission. "If they had missed that little clearing, they would have ended

Mike Johnston risked his own life to make the jump onto the side of Mont Jacques Cartier. He was once listed in The Guinness Book of World Records *for a group parachute jump onto the North Pole. (Courtesy of Mike Johnston)*

up far down the mountain. They were lucky, really lucky." Yet both Johnston and Beattie downplayed their heroism. Because the lives of strangers were at stake, they risked their own.

In order to assess wind speed for their descent, the jumpers tossed streamers from their plane each time it passed over the mountain peak. Several of these flew far off. Others disappeared as soon as they went into cloud. Finally there was a brief clear spot above the summit, and Beattie and Johnston decided to go. They plummetted from the plane at an altitude of 500 metres, well over a kilometre away from the hill. The high winds, they hoped, would take them where they wanted to go.

"Almost as soon as my chute opened, I went into cloud," Beattie said. "I lost Mike, and I wasn't sure where he was. But by the time I came out and could see again, I saw him hit and then get dragged by his parachute. When he stopped, he just lay there. I didn't exactly know what to expect."

Then, like a man being pitched from a runaway horse, Paul Beattie slammed onto the rocky ground. As he hit, the screaming wind kept his parachute inflated and dragged him through a field of boulders, some as big as a car. As he tumbled, Beattie's head, arms, shoulders and legs smashed into solid stone.

"We rode it pretty hot to get in there," Mike Johnston continued, "and we hit pretty hard. What had looked like fairly level terrain from the air, and through cloud, was really covered with those boulders.

"When I finally got stopped and got my canopy collapsed, I stood up and looked around for Paul. He landed after I did and we were both dragged toward this cliff. I managed to stop about twenty metres short of the edge, but Paul was even closer than that. I remember watching him get into a push-up position and he gave me a 'thumbs up' sign. But then he collapsed. I thought that I had another casualty to look after, so I ran over and asked if he was okay.

"When he finally caught his breath and his head cleared, he said he had bounced off so many rocks he

Paul Beattie jumped into the site of the plane crash.
(Courtesy of Paul Beattie)

thought he was in a spaceship plowing through flying boulders. 'It was like *Star Wars*,' he told me, 'like being dragged through a meteor shower.' Later on, both of us turned black and blue. There was no place we weren't bruised."

As soon as the jumpers had secured their gear, they prepared to look for Godin's plane. "From where we stood, though, we were not sure where the crash was," Beattie explained. "We radioed up to the Buffalo to get directions. They wagged their wings over the site, and it seemed they were right over us. We were already quite close to the crash, although getting down to it was tough."

Much of the upper part of the mountain was covered in a thick, dense kind of bramble bush. Beattie and Johnston struggled through this, and half-walked, half-stumbled downward.

"I was a bit ahead of Mike," Beattie explained, "when I came out on top of this fairly large rock outcrop. When I got there, I saw the crash right in front of me.

"The plane seemed reasonably intact, but I knew it would never fly again. Then I noticed the seats, and they had this brown and white kind of leopard skin pattern on them. From where I stood, the brown looked like blood, and I sure wondered what I'd gotten myself into. I couldn't see any people, but I told myself that this was what I was trained for, so I climbed down to the crash.

"The three people were over to one side, and they had dragged a tent over there, and used it for cover. It was not set up, probably because it would have been very difficult to ever get a tent pole down. Then I saw the pilot, and I know I scared him. They had seen planes going over, of course, but they had not seen us jump, and here I was standing there. The poor guy thought I had climbed up the mountain to them."

"The three kids had looked after themselves pretty well," Mike Johnston said. "One . . . young woman was the worst of the three. She had been in the front seat and when the plane hit the hill, the rock was on her side. Her right leg was broken, we figured, and as it turned out we were correct. It was badly swollen and discoloured, and it was obvious that she was in a lot of pain. We made her as comfortable as we could, and I started an intravenous line. Paul put a splint on her leg and we watched her pretty carefully."

But then the young woman, Monique, went into what search and rescue people call rescue shock. Until the SAR Techs got there, she fought to maintain her health, to stay alert — to stay alive. But once her rescuers had arrived and started working on her, she felt she was in good hands, and would now be okay. She relaxed, and in doing so actually became worse. "We had to really watch her and encourage her," Paul Beattie recalled, "because she got a lot worse rather quickly, and could have died there.

"I was pretty busy doing my thing," Beattie added, "so I really did not get a good look at their plane. I remember how young the pilot looked. He was really only a boy."

The survivors were now ready to be removed from

the mountain. However, a chopper could not hover immediately above the crash site because of the winds, so Monique had to be carried farther down the slope to a place where she could be hoisted. Christian and Edith were able to walk down to the rescue site.

A little over three hours after the searchers reached the crash site, the victims of the Cessna crash were in a hospital emergency room in Rimouski.

CHAPTER SIX

Capsized!

"I closed my eyes then, because they were really starting to burn, but I suppose, in a way, I might have been afraid of what else I might see."

Kevin DeGroot had been in the United States Coast Guard for over a decade, helping to save lives as a member of a lifeboat crew, then as an aviation technician. He was also a swimming instructor. But on May 27, 1996, Kevin DeGroot became a hero because he got lost.

DeGroot was driving to a restaurant in Pensacola, Florida, when he made a wrong turn and ended up on a pier at the waterfront. His wife Lisa, an airplane pilot, suggested they try another route because this was obviously not where they wanted to be. Kevin agreed, and slipped the car into reverse. It was noon on Memorial Day — a day the couple would not forget.

Suddenly, a stone's throw away, they saw a man jumping up and down, screaming. He seemed to be yelling to them, waving his arms and pointing toward the water. His cries were loud, frantic, desperate. His eyes were pleading.

"Something's wrong!" Lisa cried.

Kevin DeGroot stopped, jumped from the car, ran over to the man and looked in the direction he was pointing. Lisa followed, a step or two behind. Then Kevin saw people in the water.

Three metres below was a capsized boat, with people thrashing wildly around it. Most were children, and none seemed able to swim. An older man was sitting on the upturned hull, his head in his hands, apparently oblivious to the panic around him, unable to respond to the cries for help.

A short distance away a bystander was already in the water, clutching a little boy in one hand while gripping a piling with the other. The man with the child yelled for help and Kevin ran to him. Lisa went the opposite way, called encouragement to a woman holding onto the bottom of the boat, and asked her how many people had been on board.

In all there had been twelve: four adults and eight children. Already, three of the young people had disappeared.

When he got to the man holding the boy, Kevin climbed out onto a second piling and, reaching as far as he could, was able to grasp the child by the strap of his life jacket. But as he lifted the boy, who looked to be about two, the jacket tore away and the child fell back into the water. The man on the piling made a second rescue, and this time Kevin was able to haul the boy to safety.

But now Lisa was calling.

"I ran back to her," Kevin said, "and that was when she told me three kids were missing. She thought they must be under the boat."

DeGroot kicked off his shoes and plunged into the water. Then he swam to the bow of the boat, a five-metre-long vessel called the *Catfish*. A woman standing nearby had just told him she was a trained Emergency Medical Technician. She said if he was able to locate any of the children, and could get them onto the dock, she would do her best to resuscitate them.

The stern of the boat was about a metre under water. "I dived underneath, with my eyes open, and swam up into the bow and looked around in there. The water was pretty filthy, and I saw parts of dead fish and other stuff, but no children. I closed my eyes then, because they were really starting to burn, but I suppose, in a way, I might have been afraid of what else I might see."

While Kevin was searching, the woman holding onto the boat was crying out in terror, "My baby. My baby. Save my baby!" Her son, Jarrad Brazier, had been sitting on her lap just before the *Catfish* flipped over. Initially both mother and son had been trapped under the boat, but in the struggle to free herself, Judy Brazier had lost her grip on Jarrad.

"I dived back down," Kevin continued, "and systematically searched from the bow to the stern, but apart from a lot of ropes and junk, I found nothing. By my third dive, I had pretty well decided that there was no one there, but on the fourth found that wasn't so. I was feeling around closer to the back and I touched a face."

At first the thought that he had put his hand on the face of a corpse gave DeGroot a shock, but he willed himself to keep going. His hand slid across the face, then to what felt like cloth.

"I was able to grab someone's shirt," he explained,

"and I pulled it toward me. The person wearing it was a little five-year-old girl, but I didn't know that at the time. I just held onto her and then I had to come up for air. But I couldn't get her to the surface with me. At this point I realized she was tangled in the boat lines, and I couldn't get her free. I went back under the boat to try to untangle them."

DeGroot worked frantically to free the child, while holding his breath as long as he could. He would grip tightly, surface for a gulp of air, then go down again. "I finally got her down and out from under the boat," he recalled, "and then I had to surface again."

This time DeGroot succeeded in getting the child's face above water, but when he noticed foam coming from her nose and mouth he realized she was probably dead. He began to tow her to the sea wall, but soon found he could not pull her farther. A line from the boat was still wound around her legs, keeping her tied in the water.

"I finally got [her legs] untangled," he continued, "and then I was able to bring her over to the wall. Somebody lowered a set of car battery cables to me, and I wound them around my arm and got pulled far enough up so that I was able to hand the child to somebody. I dropped back into the water because the people on the shore wanted to tie off the boat so that it would not drift away. At this stage I was pretty exhausted and I knew I could not go under it again."

On the dock the woman who had spoken to Kevin immediately began to perform CPR (cardio-pulmonary resuscitation) on the girl. She had stopped breathing, so the task was daunting. However, Deanna Adams refused to give up. The child, Keidra Sanders, finally begin to stir, then to breathe. By the time an ambulance arrived, she seemed to be out of danger. She was rushed to nearby Sacred Heart Hospital.

In the meantime, Kevin DeGroot was still in the water. "Somebody threw me a line," he said, "and I tied the boat off. But that caused me trouble."

A crowd had gathered, and in their haste to secure the *Catfish*, did so too hurriedly. As soon as Kevin had the line tied to the boat they pulled it sharply toward the dock, forgetting that he was trapped between it and the steel-ribbed pier. The bow of the boat slammed into the pit of his stomach and knocked his wind out.

The move left him gasping for breath, and in his exhausted state it not only hurt him, but could have easily drowned him. Now DeGroot's extensive rescue training once again came to his aid. He tilted his head back, continued to tread water, and looked for something to grab. He struggled over to a life jacket floating nearby.

The jacket was from the *Catfish*. It had drifted away when the boat capsized, bringing an end to what should have been a fun cruise for two families, the Braziers and the Sanders, and one or two others. A man named S.T. McCreary owned the *Catfish*. He had agreed to take the people for a short pleasure trip into Pensacola Bay, but the waves there were too choppy, so the group came back to land. They were about to tie up when their troubles started.

As McCreary eased his boat toward a sheltered

docking area, he cut the motor and let the *Catfish* glide up beside a wooden piling. Then he tossed a rope toward the piling to secure the boat. Unfortunately the toss missed, and the boat drifted to one side, too far away for a second throw. When McCreary went to reposition the boat the engine would not start, even after repeated attempts. The *Catfish* was dead in the water. A wave washed over the stern, then another and another. As McCreary frantically tried to start the engine, one of the men on board began bailing desperately. His efforts were not enough.

One particularly large wave flipped the boat upside down. Everyone on board ended up in the water. McCreary pulled himself onto the upturned hull, Judy Brazier clung to the side of it, and the other two adults got to the breakwall. All eight children remained in the water, where they were when Kevin and Lisa DeGroot arrived.

"I remember grabbing the life jacket," Kevin said, "and then I noticed a woman half out of the water, holding onto what looked like an old electrical cable sticking out of the wall. So I swam over to her and tried to get her to take hold of the life jacket, but she was too scared. She wouldn't let go of the cable."

The more DeGroot tried to convince the woman to take the jacket, the more she resisted. He knew her arms were tiring and she would not be able to grip the cable for long. Fearing she would drown if she relaxed at all, he stuffed the life jacket around her, under her arms, and hoped it would stay there. Then he swam to another piling and caught his breath.

Meanwhile, on the dock, the crowd of onlookers

Lisa and Kevin DeGroot saved the lives of Keidra Sanders and others when the Catfish capsized. (Courtesy of Lisa DeGroot)

had grown. A woman rushed to help Lisa, who had noticed that one of the children had climbed onto the boat hull, alongside McCreary. After a great deal of effort the two women plucked the child from the hull.

A man named Michael Patterson, who had been fishing from the pier when the whole episode began, also came to assist. He clung to the sea wall, then reached over as far as he could and retrieved one child. Then, realizing that McCreary would be too heavy to lift, he sprinted to a nearby building and called the police.

By now Kevin DeGroot had been in the water for some time, and Lisa feared for his safety unless he returned to the dock. She knew he was close to exhaustion because of the dives under the *Catfish*.

"My wife yelled at me and told me to get out," Kevin recalled, "and I knew she was right. I was so tired that I had to have her help me up onto the sea wall. She told me they used ropes to save some of the others."

Soon the police arrived. Andy McCoy was one of the first officers there. He immediately jumped into the water and began searching for two of the children who were still missing. Because he was an experienced swimmer, McCoy dived under the Catfish several times over the next 20 minutes, but was unable to find anyone. He then assisted others in securing heavy ropes to the boat so that it could be lifted out of the water in order to make searching easier.

That's when Andrew Jackson was found. Fifteen-year-old Andrew, a non-swimmer, was trapped under the boat, his ankles encircled by a rope. McCoy helped lift him into a marine patrol vessel which came to the scene, and then attempted to resuscitate the boy. Unfortunately, all his efforts were in vain. The same afternoon a police diver found the body of four-year-old Jarrad Brazier, the child who had slipped from his mother's grasp when they were trapped under the *Catfish*.

Keidra Sanders recovered completely, even though she had been under water for at least 15 minutes when Kevin DeGroot found her. No wonder she is so grateful to the young Coast Guardsman who got lost on his way to a restaurant.

He will always be her hero.

CHAPTER SEVEN

The Bottomless Pit

As the remains of his house slid out from under him,
he dived for the door. In that instant the house was gone.

It was April, 1971, near St. Jean Vianney, Quebec, 16 kilometres from Chicoutimi. Peter Blackburn walked into a field by his barn to see what was bothering his cows. They had been grazing quietly until all at once they began to shy away from a particular spot. Then they stampeded to a far fence and remained there, obviously agitated.

Blackburn had walked less than a hundred metres when he saw something in front of him. In what had been a flat, ordinary-looking pasture was a gaping hole in the ground, three or four metres across, and so deep he couldn't see the bottom.

He stepped back, bewildered. Then he moved forward and peered into the pit with alarm. He was concerned that his cattle might stumble into the hole before he could get them back to the barn, so he decided he would see to the cattle, and worry about the cause of the pit later.

After rescuing the cattle, Blackburn reported the hole in the field to local officials, who contacted provincial authorities. They came up with several theories on what might have caused the mysterious pit, but came to no firm conclusions. All anybody knew for sure was that it was a sink hole of some kind, at least 50 metres deep and perhaps much deeper. The junk they threw into it took a long, long time to hit bottom.

On May 4 of that same year, most Quebeckers were watching the NHL playoffs, cheering for the Canadiens as they battled the Blackhawks.

The May 4 game was hard-fought playoff hockey, with the winner not decided until the final second. The game was still going on at 11 o'clock that night, and much of the population in the town of St. Jean Vianney, near Peter Blackburn's farm, was tuned in. The town of 1300 was a comfortable bedroom community like many others across the province.

Residents of the town suddenly felt their houses shake, at first a little, then more strongly. Their TVs flickered and chandeliers swung. Dogs yelped at the sound of things crashing outside. People ran out and stared in disbelief at the pit of doom that was their town.

Entire houses were moving, breaking apart, falling in on themselves and disappearing. Parked cars and trucks disappeared too, as roads and driveways vanished. Street lamps swung, crashed to the ground and exploded in showers of sparks.

There were screams — wrenching cries of terror out of the darkness, calls for help, sobs, moans and half-heard prayers in the night. Cries of the young and the old who were dying in terrible ways, just as their town was dying around them.

No one knew what was happening.

Jean-Paul Boivin dashed outside, alarmed, as his home started shaking. Dozens of screaming people in nightclothes stood in shock watching a neighbour's house break apart and disappear.

Boivin ran back inside and led his wife and their four children out of the house. Less than a hundred metres down the street he glanced back, in time to see their home sink from sight.

A block away, a man getting ready for bed heard a rumbling outside. His house began to tremble and he went to investigate. As he walked toward the street his house vanished behind him, taking his wife and their 15-day-old daughter, who had been baptized only hours before. He never saw them again.

Roger Landry was in his basement when he felt the cement floor shaking under his feet. He immediately raced back up the stairs, to find that his house was already half destroyed. He frantically climbed through what had been the kitchen, screamed for his wife, and in a blind panic tore away rubble leading to their bedroom.

Seconds later, as the wreckage slid out from under him, he dived for the door. His wife and their five children, Jeannette, Hélène, Anna, Denis and Bruno, all died that night.

"I called them until I thought my lungs would burst," he said later, "but there was no answer. It was dreadful, indescribable." Landry searched for his family for hours in the darkness. Finally, desolate, he took shelter with relatives.

Another family survived the catastrophe through

dogged persistence, courage and a mother's love.

Mrs. Marcel Boily was in the basement, too, doing laundry. As she carried a bundle of clean clothes back up, the stairs shook and she heard loud noises outside. She dropped the clothes and went to the door. People outside were screaming and crying, yelling that the street was collapsing and urging the neighbours to run. Boily left the door open and ran to waken her husband and children. "We raced up Harvey Street in our bare feet," she told *The Montreal Star*. "People were screaming frantically and shoving each other as the ground kept shaking. I saw one person fall down and disappear."

When the family was at a safe distance she noticed that her 12-year-old son, who had been sick in bed all day, was missing. She told the rest to stay put and ran back down the street. A policeman tried to keep her from returning to rescue the boy.

"I pushed him away and ran, but he caught up to me and grabbed me. I struggled with him, and tried to tell him about my son, but he wouldn't listen, so I began hitting him," she said.

The two struggled and fell to the ground, but Boily managed to free herself and run off. As she got to her house she heard a loud roar and watched, amazed, as her neighbours' house slid away.

"I was so scared, I couldn't remember at first where my son's room was," she said. She picked him up in her arms and carried him outside. By the time she reached the end of the driveway the whole house was gone.

As she stood staring at the gaping hole she felt a tug on her arm. It was the same police officer who had tried to stop her. Together they ran to safety and joined the rest of her family.

In another area of St. Jean Vianney a bus was carrying workers home from a late shift at the Arvida aluminum plant. The rain was pouring down so hard, visibility was extremely poor. Driver Jules Girard was proceeding slowly, hunched over the wheel as he tried to see through sheets of rain. The windshield wipers slapped back and forth, but they didn't help a lot. The headlights reflected on the rain and made it even harder to see much. But then Girard realized he wasn't seeing *anything*.

His lights shone out into space. There was no white line, no shoulder. No road at all!

He slammed on the brakes. Behind him, half a dozen lunch buckets clattered to the floor. The passengers lurched forward in their seats. All conversation stopped.

For a few seconds no one moved. All eyes stared forward, at the spot where there was — nothing.

Girard peered out into the darkness. Then he opened the door, intending to check the road. Suddenly the front of the bus dropped.

People scrambled to the back of the bus. Somebody opened the emergency door and the passengers clambered outside. Girard was the last to leave.

Seconds later the earth collapsed and the bus tumbled, end over end, out of sight.

Shortly after midnight a telephone rang at the Canadian Forces Base at Bagotville, 45 kilometres to the southeast of St. Jean Vianney. The information relayed was jumbled and incomplete. What did get through was barely believable. The caller said that a town was disappearing and a lot of people were dying. Help was needed urgently. There were few other details.

"When we got the message from RCC we didn't know what to think," recalled Mel Furlotte, a flight engineer with Bagotville Search and Rescue at the time. "They told us virtually nothing, but to have a helicopter ready to go right away, and to put a couple of extra stretchers in it. We had no idea where we were going, or why.

"By the time the helicopter left Bagotville, further messages that had come in mentioned a landslide in a little town called St. Jean Vianney. None of us knew quite where it was, but we knew it had to be to the north, somewhere around the Shipshaw Dam. We'd seen the dam from the air on lots of occasions. In the meantime, the weather is terrible. There's fog, rain and high winds. We're in the most useless helicopter in the world, and we don't know where we're going. Fun, eh?"

The helicopter in question was the Vertol H-21, the so-called "flying banana" because of its long curved shape. It was not a dependable aircraft: difficult to fly, underpowered, prone to stalling. A few months earlier Furlotte and his crew had crashed in an identical chopper because of a power failure at a critical time.

"So none of us were super confident about things," Furlotte added, "but luckily the guys up front were the best in the business." The "guys up front" were the pilots, Palmer "Tiny" Wenaas and Jack Farncombe.

"We followed the highway up past Arvida," Furlotte continued, "and you could hardly see a thing as we went along. Then we found the Shipshaw Dam somehow and figured where we were going was somewhere close by. There were no lights though, and we didn't know then that all the power was out in St. Jean Vianney.

"Finally we saw some flashing lights from police cars, so we landed and asked what was going on. They told us that right out in front of us there were no roads, no houses, no trees or anything else. They said there had been a landslide and everything had gone down into it.

A Vertol H-21 helicopter hovers over a half-buried house. (Manny Soberall/Canadian Forces photo, courtesy of Mel Furlotte)

This house is one of the many that slid into the bottom of the enormous sink hole. (Manny Soberall/Canadian Forces photo, courtesy of Mel Furlotte)

"So we took off and found the hole.

"It was the most monstrous thing I had ever seen. We could see houses falling over the side, smashing up, being sucked down and being totally destroyed. There was lumber, bathtubs, cars, the sides of buildings — you name it — floating away. It was absolutely stupendous. And we didn't know then how big it was.

"Tiny decided to fly around the circumference to see if we could find anyone alive down inside. At this time we had no SAR Techs on board, only Ron Servos and I in the back. Ron was at the front right door, and I was at the back, looking out the left side.

"We had only a single light, a landing light, but it was not much good. The thing was a piece of junk, and it would only work for about five minutes at a time, then it would blow a relay switch and go out. Every time that happened, we would fly out of the hole so that at least we were over firm ground if the engine quit.

"We went back and forth, down in the pit, trying to locate anyone who might have been alive. The whole thing was so eerie. In the darkness and rain, you really could not tell how awesome it was. We would come over the edge, then go down and hover, just above the moving mud. None of us had a lot of faith in the helicopter, and we knew if we stalled we would never get out. There just wasn't anyone alive down there that we could see. It was a terrible place. We kept searching until about three-thirty and found no one. By then we needed gas so we decided to go back to the base and wait for morning so that we could see. I remember being exhausted by the time we left."

As soon as the H-21 touched down back in Bagotville the ground crew checked it over and prepared it to set off again when the operation resumed at dawn. The four fliers had time for only a short rest. By 6:30 they were back in the aircraft, awaiting takeoff clearance, accompanied by para-rescue specialist Rod Verchere. Verchere was known as a totally fearless man, particularly in his para-rescue work. He was in the right place that day.

"He is probably the most dedicated person I've ever

Rod Verchere risked his life, over and over, to search for survivors in homes that had slid into the gigantic sink hole in 1971.

met," said Mel Furlotte. "He's afraid of nothing, and would do anything, anywhere, if he felt there was even a remote chance of saving a life. He went into situations that were terrifying, and operated with a complete disregard for his own safety. I know if I was in trouble, I would be thankful to see him. That was why we were lucky to have him with us in the pit."

The chopper took off and followed the Saguenay River north, toward the doomed community. Rain was still drizzling down and the clouds were dark as the men flew over the placid river. The flow was clear and clean, until they got near the location where the Rivière aux Vases joined it.

"I remember looking over one of the pilot's shoulders," recalled Furlotte, "and noticing how the river suddenly became very muddy. Then a bit farther along, it was clear again. That's when we noticed a stream of

Mel Furlotte was instrumental in rescuing survivors from the St. Jean Vianney sink hole.

silt coming in from the right side, with all this junk floating out on it. Until that moment, we had not connected the big pit at St. Jean Vianney with this runoff into the Saguenay. But there it was."

What had been the tiny Rivière aux Vases was now a 30-metre-wide mass of brown water and mud, full of debris from the enormous sink hole.

The rescue helicopter had just turned inland from the Saguenay when they spotted a corpse trapped in the muck. "We went down," said Furlotte, "and then swung in over the body, a man. By the time we got him out of there and had the body in a bag, we had mud to our hips. The stuff was like thick, grey soup that moved."

The body was taken to a nearby emergency measures centre. The chopper went on toward the sinkhole itself.

Lisette Lepine spent a terrifying night atop her sinking car, calling for rescue. (The Montreal Gazette)

"When I first saw it, I couldn't comprehend the enormity," said Verchere. "It was so vast, and so unbelievable, I thought there would be hundreds dead down there. It was . . . about a half mile [800 metres] across, and two hundred feet [60 metres] deep, and there were lots of houses and parts of houses down inside. Some others were still falling off the cliff, and this was six or seven hours after the slide started."

Almost as soon as they arrived on the scene, the crew located a car near the bottom of the sink hole, but not far from one side. A woman clung desperately to the roof, knowing that at any moment the wreck could shift and she would be sucked downward to her death. Would-be rescuers stood by the edge of the pit helplessly.

"We lifted her out of there," recalled Furlotte. "Rod was in a wetsuit and he got a horse collar around her. She had been there all night, yelling for help, and when we arrived she was shaking and crying, but could not speak. Her voice had completely gone. She had a few bruises, but was in fairly good shape otherwise. We learned later than her name was Lisette Lepine."

The men in the H-21 now began to search the hole for survivors. They began with the wrecked house that lay deep inside the pit.

"We would hover over a house," explained Furlotte, "and if the pilots could get one of the back wheels on something solid — a roof, or a porch, or something like that — Rod would hop out and go inside. In all, he checked seven or eight, but we found no one. When he came out, we would haul him up."

It wasn't easy to get through the shattered buildings. But despite the jammed doors that had to be kicked open, and ruined walls that needed climbing over, Verchere searched every room. He made his way through unfamiliar rooms littered with upturned beds, refrigerators, sofas and the remains of televisions piled together, in the dimmest of light. Making his way through all the junk, over broken glass, took time. And still the buildings were moving, their roofs and walls collapsing around him.

"Rod would be gone for the longest time," Furlotte said, "but then he'd come out; we would pull him up and take him to the next. A couple of times, he wasn't coming out and the house was falling apart, and we would be sure he was gone, that he was dead for sure. Then he'd pop out, give a thumbs up and we'd hoist.

"During all this time we were more scared for him than he was. Finally we told him 'Enough. No more.' Then, 'In a minute,' he'd say. 'Let me down over there. I want to check that one.' He kept doing that until we had to leave for fuel. When we got back there were no more houses."

"We then looked for cars, things like that," Verchere said. "I'd get hoisted down in order to check each one we found. They were all empty. After that, it was mostly picking up bodies."

They worked until late in the afternoon, when the primary recovery work was done. Then Wenaas put the chopper down near a police command post. Finally, the exhausted rescuers could rest.

"I had been handling bodies," Verchere said. "I was covered in mud, and I was so beat I could hardly move."

❖ ❖ ❖

For several days cleanup continued at St. Jean Vianney and along the Rivière aux Vases and the

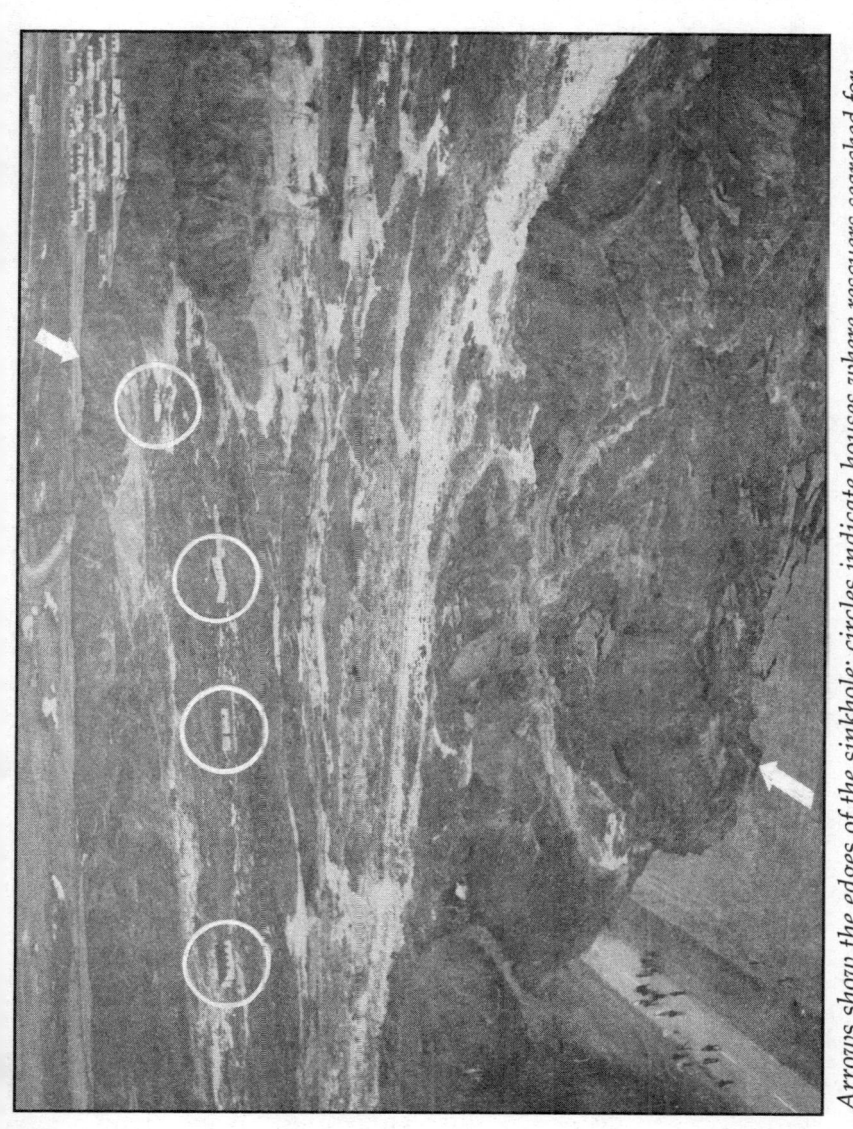

Arrows show the edges of the sinkhole; circles indicate houses where rescuers searched for more survivors. (The Montreal Gazette)

Saguenay. Thirty-one people were killed in the catastrophe, fourteen of them children. Thirty-six homes were completely destroyed, as were several cars, one bus, and an unknown number of family pets. Almost 2000 were evacuated to safer places and some 15 million tonnes of mud and earth were displaced. And the cause of the disaster? The experts blamed several days of rain, a possible underground stream, unstable clay and shifting water-soaked soil that had been there since the last Ice Age.

Rod Verchere's action during the St. Jean Vianney rescue won him the Star of Courage, Canada's second-highest award for bravery.

CHAPTER EIGHT

Between Sea and Rock

The man was now bobbing around on top of
house-high waves that were crashing on the rocky shore,
breaking up, then falling back on themselves.
His chances of survival were slim.

As a little boy in Clearwater, Florida, Tyler Jennings used to watch search and rescue helicopters flying over his house. He'd hear the throb of their engines, watch their size, their speed and their ability to actually stop in the air. One day, he vowed, he would ride in a chopper like that. It would be a dream beyond anything on earth.

Years later Tyler made his dream come true. When he was 21 he drove to a recruiting centre in Mobile, Alabama, and joined the United States Coast Guard.

After a tough and stressful boot camp at Cape May, New Jersey, Jennings was posted to a Coast Guard ship, the cutter *Tampa*, as a seaman. Later on he did flood relief duties on the Mississippi River, prior to being sent to Sitka, Alaska, in the winter of 1994. He was still there in late December 1997, as a flight mechanic on search and rescue helicopters.

The Coast Guard presence is important in Alaska, the state often called America's Last Frontier. Some 1800 active-duty members serve there, and search and rescue is a vital component of their work. In 1997 alone there were 1140 missions conducted and 331 lives saved. And while equipment for search and rescue is located at various places, the twelve helicopters are based at two: Kodiak and Sitka.

Sitka is on Baranof Island, on the outer coast of Alaska's Inside Passage. Like most communities there, the town is accessible only by air and sea. Just before Christmas in 1997, Tim Keeler was headed there with a boatload of fish.

"He had been fishing in the coastal waters," explained Tyler Jennings, "and the cargo was to be sold in Sitka. On the way there he ran into bad weather, and even though there were places he could have gone and been safe, he still decided to try to get to Sitka. Unfortunately, that was a bad call."

Keeler's boat was the *Callisto*, a 12-metre-long diesel-powered trawler. When the Gulf of Alaska storm hit him, the main engine on board quit, and there was not enough battery power to restart it. The craft drifted helplessly on the Pacific rollers. They drove him, at first slowly, but then faster and faster, toward the wild, cliff-edged coast of Baranof Island. He still thought he could get under way again, and he only called the Coast Guard as a precautionary measure.

Keeler fought the wind and the waves for hours, alone, in darkness and rain, but still his boat was being blown toward the cliffs and he could not get the engine going. Just after ten at night on December 22, the Juneau, Alaska, Rescue Control Center sent a helicopter from Sitka to assist him.

The helicopter attempted the rescue, but they finally had to leave because their hoist cable became frayed during the numerous attempts to reach Keeler. That's when Tyler Jennings was called in.

"That night I had a weird feeling that I would be called out," he said. "I was so sure, I even put our portable phone at the head of the bed. Ordinarily, I never did that. But sure enough, at twelve-thirty the phone rang. . . . When I got in I went into Maintenance to see what was going on. I had just opened the door and my Operations Officer came bouncing down the stairs. He told me the helicopter that was on scene was coming back because they couldn't get the guy. He said we were going out right away."

With that Jennings and the rest of his crew began preparing their H-60 Jayhawk for the mission. This type of helicopter is almost 20 metres long, rotor blades included. It can fly for six to seven hours at a normal cruising speed of about 140 knots (over 260 kilometres per hour). It has two engines, carries a crew of four, and is both versatile and dependable. It cannot land on water.

"It was about one-thirty a.m. when we took off," Jennings recalled. "I remember checking the weather, and it was not too bad near the base, but after we had flown for a few minutes we made a left turn out over the Gulf of Alaska. That was when my world as I knew it came to an end."

As soon as the Jayhawk cleared the cliffs bordering the sea, it faced the direct assault of the Alaskan gale. The shrieking wind tore in from the ocean, blowing freezing rain sidewise. Outside the helicopter the low cloud ceiling made the night absolutely black, and the plane was tossed every which way in the storm.

"The wind was ripping so hard that the sleet and snow made it impossible to see anything, even when we wore night vision goggles," Jennings said. "We were under eight hundred feet [240 metres], and even then I could barely see the ocean. Sometimes there were faint traces of waves crashing down there, but that was all."

"When we were on our way out, we met the other helicopter coming back," he continued. "They briefed us on the weather and described the situation. They told us it would be tough to hover with the sea state and the pressure changes because the cliffs were close to the guy."

Jennings and the other men in the relieving helicopter, pilots Dan Molthen and Doug Taylor, and rescue swimmer Mike Fish, also learned that Tim Keeler was no longer on board his boat. During the time the first helicopter had been overhead he had transferred to a canopy-covered life raft, and his vessel, the *Callisto*, had sunk. The man was now bobbing around on top of house-high waves that were crashing on the rocky shore, breaking up, then falling back on themselves. Sooner or later one of them could toss him onto the rocks and tear him apart. If he was not rescued right away, his chances of survival were slim.

"When we got there the first thing we did was a visual approach around him to see what was going on and to make sure he was still in the life raft," said Jennings. "I had the door open and I had completed my rescue checklist."

The Coast Guard often uses a hoist, payed out from a hovering helicopter, to rescue people stranded in the water. This training session shows much calmer conditions than those faced by Jennings and his crew. (U.S. Coast Guard)

The helicopter crew had their Night Sun spotlight on, and could pick out the life raft in the surf below. Had Keeler tried to ride the waves to shore, "he would have been beaten to death," Jennings explained. "There was no beach to land on, there were sheer cliffs up from the water's edge. And the seas were so high there was no set pattern to them. They were rolling in, then coming back out. I had never seen anything like it."

Tim Keeler's situation was terrifying. Riding the raft was precarious in itself, but he also had to contend with the driving rain, the roar of the wind and surf, and now with the racket of the helicopter overhead. At times the machine would be right in front of him, then a couple of seconds later it would be blown away to the side. Then it would be far overhead, then drop so close he feared it would crash. Flying into the teeth of a gale was never easy, he knew, and being rescued on this wild night might be impossible. He realized he soon could die.

"He was close to the cliffs," continued Jennings, "and because of that, air pressure changes really affected us. The helicopter was being blown violently all over the sky. When I saw the guy, he was standing in the [opening] of the raft canopy, waving to us, but I knew that getting a rescue basket down to him would be tough. There was a real downdraft and swirling wind off the mountains.

"We tried our first hoist from about forty feet [12 metres], but the next thing I knew I was yelling into the radio, 'Up! Up! Up!' We had been blown down very close to the water. We tried again from forty, but the same thing happened. On the third hoist, we tried at seventy feet [over 20 metres], but got blown straight down to twenty [6 metres]. The wind was about fifty knots [90 kilometres per hour] and it blew us back toward the cliffs. That was a problem."

The helicopter crew intended to use the rescue basket to retrieve Keeler. Swimmer Mike Fish volunteered to go down into the churning sea to try to save the stranded fisherman, but Jennings talked him out of it.

"As flight mech, I said I did not want to put him down," Jennings recalled, "and the pilots agreed with me. I sure would have hated to have two people in the water. . . . If Mike had gone down, he might have gone to certain death, because he would have been smashed up by the sea."

So the rescue basket was the last resort.

In order to operate it Jennings hooked himself into the on-board safety harness and, wearing kneepads, knelt in the open doorway of the Jayhawk. Then, with one hand on the cable and the other on the hoist control, they attempted to drop the rescue basket as close to Keeler as possible. To make this work, the two pilots had to do some of the finest flying of their careers. For his part Mike Fish shone the Night Sun into the dark cliffside, to be sure the tail rotor blades did not contact the nearby rock face. Fish also hovered at Jennings's side, and assisted in calling the radio commands for position.

"There were times when I tried to hoist," said Jennings, "and we had the rescue basket down and then we would get a severe downdraft. The basket would hit the water and the pilots would be correcting, and the cable would be ripped right out of my hands. There was often no way I could hang onto it.

"Then the cable would come down, hit the bottom of the door frame, or because of the wind direction, would be jerked up and to the left, out of my hands, when the pilots had to pull up. Then, as soon as we got up, the basket would swing wildly, away out underneath the rotor path."

During all this Jennings was being sapped of his strength. Yet he held on, his hands raw and freezing from the constant cold rain and sleet that drenched him and blew into the aircraft itself. After a while he had so little feeling left in his hands he had trouble grasping the cable.

Then another problem arose.

Neither Molthen nor Taylor was able to get his bearings in order to try to stabilize the Jayhawk. Fish and Jennings readied two flares and tossed these out the door. The markers were momentarily visible, but then were washed from sight by the choppy sea. The pilots brought the helicopter higher and regrouped, and three more flares were dropped. This time the visual reference was a little better.

"We did six hoists in all," Jennings recalled, "and the first five were unsuccessful. But on the fifth, I simply dropped the basket into the water, and when that stopped the swing, I figured if we could somehow drag it over to the guy, we might be able to make things work."

All the while, Keeler was gripping the opening in

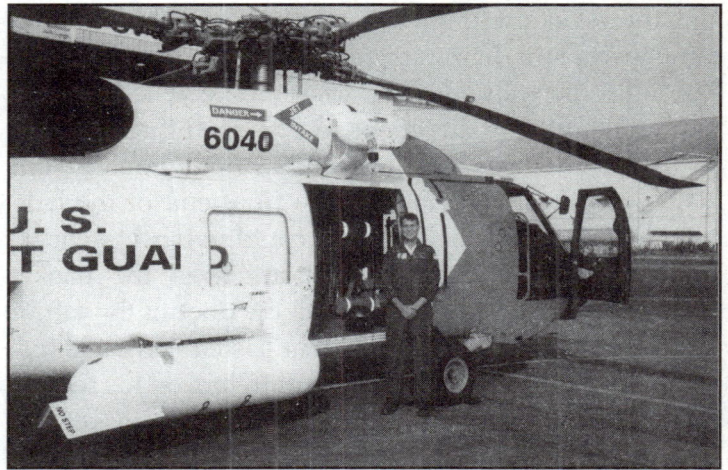

Tyler Jennings at the bay door of a U.S. Coast Guard helicopter. (Courtesy of Tyler Jennings)

the canopy of the life raft, holding on for all he was worth. He was bitterly cold now, tired and fighting seasickness because of the violent lurching of the raft. But he knew he had to stay alert in order to be ready to somehow get into that basket if the chopper crew could get it to him. Often, just when he thought it was coming near enough, the basket would jerk out of the water and swing far out into the night. Then he would watch it through the sheets of freezing rain and fight the fear of being abandoned and thrown to his death against that ever-nearer wall of black rock.

In the meantime the Jayhawk went higher, turned, then lined up behind the raft. As Jennings struggled to maintain a sight line, he relayed directional commands to the flight deck. The helicopter came lower, slowly, and the hoist cable snaked away. When the chopper came lower still, Jennings payed out the cable until the basket hit the water. As soon as it did he yelled into his radio mike to give directions to approach their target.

"The basket was back near the tail area," he said, "but it was going forward toward the front end of the helicopter. When it got out as far as it could swing, I [secured] the hoist in the DOWN position. Then the basket streamed out in front of us and hit the door of the raft."

In one desperate motion that lasted perhaps five seconds, Tim Keeler let go of the raft, braced his feet on its side and dived headfirst into the rescue basket. He squirmed around into a sitting position and then held on for dear life. The basket then plunged into the sea.

"I was on the intercom saying, 'The survivor is in the water. The survivor is in the water,'" Jennings recalled. "Then the helicopter climbed up to the left, but our hoist could not get the man out of the water as fast as we were moving, and I saw the poor guy get dragged through the waves.

"I was paying out whatever slack I had at that point because after all he had been through, I didn't want him injured now. Then I started bringing him up, and he finally broke out of the water, but immediately started this huge pendulum swing. We were at about a hundred and twenty feet [36 metres] in the air now, and I had a lot of trouble trying to control the cable. I knew this guy was on the roughest ride of his life.

"We finally got him to the door and inside. Surprisingly, he was still coherent, despite what he had been through."

Commander Doug Taylor (left) and Lieutenant Dan Molthen (second from left) receive the Distiguished Flying Cross for their heroic rescue. (The Sitka Sentinel)

Tim Keeler survived his frightful ordeal, complaining only that his stomach hurt "a little bit." A doctor who examined him later that night marvelled that the stomach pain was his only discomfort. He recovered fully.

It was almost 5:30 when Tyler Jennings got home that morning. Despite being desperately tired, he was still keyed-up after the events of the night, and found himself unable to sleep. As he told his wife Cathy about the mission, he noticed that their six-year-old daughter Kimberly was listening, her eyes like saucers. When the story ended, she sat down and drew a picture of a helicopter, a line into the water, and a man with a big smile on his face. Below the drawing were the words: "My Dad — the Hero."

CHAPTER NINE

Trapped in the Wreckage

All four people were trapped in the plane.
Then they smelled gasoline.

Thérèse Fournier's second flight ever was on a stretcher in the back of a search and rescue helicopter. Her first was in a plane that crashed. That time, what had been intended as a short sightseeing ride over northern Ontario bush in a small float plane, quickly became an unforgettable nightmare.

The man flying the small Stinson 108 float plane that day, September 30, 1985, was Jean-Paul St. Jean, a man known as a careful and conscientious pilot. His sister and mother were in the back seats of the Stinson, while Thérèse Fournier sat beside him in the front. When the plane took off about six p.m., Jean-Paul told his wife Marie that he and his passengers would be back in an hour.

The sun was shining on the water at the beginning of the flight. But within five minutes the sunshine gave way to fog; fog so dense it not only covered Granite Lake, where the flight began, but completely obliterated everything else as far as the eye could see.

St. Jean lost his bearings, the engine stalled and the plane crashed.

"We were hardly in the air when the fog covered everything," Thérèse Fournier said. "We could see nothing. Everything was white. Jean-Paul told us to pull our seat belts tighter because he was afraid we were going to crash. Then we did and I blacked out."

The Stinson had travelled less than 30 kilometres before the crash. As his mother urged him, in French, to "do something," Jean-Paul did his best to re-start the engine, but to no avail. The aircraft came down in the bush, close to a natural gas cutline, roughly a kilometre from the nearest highway. The nose of the plane hit first, the wings were sheared off and the windows of the cabin shattered. Both floats were torn away, as was the right door, the one beside Thérèse Fournier. She was thrown against that door, and as it broke so did several bones down the right side of her body. One of her shattered ribs punctured a lung.

Jean-Paul went headlong into the instrument panel, and the jagged edge of a metal support beam raked his face, slicing his cheek open and exposing the teeth on the left side of his jaw. The two women in the back were thrown against the front seats, but their seat belts held. However, because they were both injured by the impact, they were in shock, silent. All four were trapped in the plane.

Then Jean-Paul smelled gasoline.

"It was a terrible feeling. None of us could escape," Fournier recalled, the tremor in her voice underscoring the seriousness of the situation.

At one time or other all four victims lost consciousness, although that in itself was almost a relief because it masked the fears they felt. When night came the darkness, the strange sounds of the forest and the knowledge that they were trapped and could be burned alive all played on their minds. The fact that marauding bears could be nearby added to the terror.

And then it started to rain.

At first it was a drizzle, then the drizzle became a downpour. "It rained so heavily, we stopped worrying about fire," Thérèse Fournier explained, "but because there was no door on my side of the plane, I was soon soaking wet."

So the long night continued — a night of endurance, prayer, pain, hope, fear and loneliness.

But not abandonment.

Back at their home on Granite Lake, Marie St. Jean worried as she waited for her husband's return. She had watched the takeoff and had stood peering at the sky until the white, brown-trimmed Stinson disappeared over the far shore. But when the unexpected blanket of fog appeared out of nowhere soon afterward, she became alarmed. She paced the shore, listened for the sound of an engine, hoping against hope that her family would soon come back. All the while, the fog thickened.

Finally, when her apprehension turned to alarm, she went into the house and called Lakeland Airways at Temagami, a few kilometres to the south. When no one there had seen the St. Jean plane, a company employee notified the local Ontario Provincial Police detachment. A while later the OPP called the Rescue Co-ordination Centre at Trenton, Ontario.

The Trenton RCC did an immediate phone check through the area where the plane might have sought safety. Initially nothing turned up, but at 9:20 that evening word arrived that a weak signal had been picked up by other planes. A Buffalo aircraft with two SAR Techs on board was sent to have a look.

"Al Houle and I went up," said veteran SAR Tech Charlie Fleming, "but by the time we were over the area, it was raining so hard and the clouds were so low, we couldn't search properly at all. Finally we had to land at North Bay and get the OPP to take us in by road. We went as far as we could that way, and then the two cops and Al and I started to walk. We had a hand-held homer, and not long afterwards we started to pick up weak ELT signals with it."

The homer is an electronic homing device, specifically designed for the type of search the four rescuers were doing. Under favourable conditions the instrument would indicate the proximity of emergency signals, so that searchers could zero in on the source. For various reasons on this particular night, conditions for

Charlie Fleming was one of the first SAR Techs to reach the crashed Stinson.

doing so were far from favourable.

"It was horrendous," Fleming recalled. "With the homer we had, the ELT signal was bouncing all over the place, off power lines and hills because there was iron in them. We would go up a logging road or a cutline, or through the bush, thinking the plane was in that direction. Then the signal would fade out or seem to come from someplace else. It was in the middle of the night and pouring rain most of the time and we were cold and wet all the time. There was no snow on the ground, so it was absolutely black in the bush. Walking through the underbrush was awful because you just couldn't see. I'll never forget it.

"But then, about six o'clock in the morning, as we were homing in on the ELT, I heard this guy talking to me on my headset. He was saying 'Mayday, Mayday, Mayday! There are four of us in the airplane and we are going to die!' Then the guy gave his name, his call sign and his location near a lake. It was the float plane we were looking for. At that stage, we had probably walked ten to fifteen miles [16 to 24 kilometres] through the bush in the dark and the rain, basically going in circles, trying to follow the signal. One of the OPP guys said he knew where the lake was, so we humped back out to the four-by-four and drove there. Then we found another cutline and started to walk again, still following the ELT."

In the middle of the night the phone rang at the home of SAR Tech Al Williams in Trenton. "It was one of our pilots, Gaston Cloutier," said Williams. "He told me about the crash, that our guys were involved in a ground search, but that they might need help. We took off at first light. Ed Holleman was the SAR Tech with me.

"We refuelled in North Bay, and then flew under the clouds to the search area. We picked up the ELT right away, and then we saw the crash. At about the same time we could see the ground searchers half a kilometre away, so we vectored them toward the site and then circled over it ourselves. We really expected the worst because there was no sign of life down there at all.

"When Charlie got to the crash, he called up to tell us there were four alive. Then I imagine he got really busy because we had no further radio contact with him. That's when Ed and I decided to jump."

Williams and Holleman donned wet suits and parachuted into a lake, because doing so seemed safer than trying to land in the bush. Once down, they swam to shore, collected their equipment and walked to the downed Stinson.

Because the four people were alive in the wreckage, the searchers began a frantic race against time to get them out. Suffering from both shock and hypothermia, none of them would live much longer unless they got to a hospital quickly. But getting the victims out proved to be extremely difficult.

"Because the plane was in the brush, several small trees had to be cut down just to get in to treat the people," said Williams. "And working on them within the confines of the plane was tough. There was just no room. We had to use a crash axe to cut away part of the plane, and then rip the front seats out to get to the two in the back. As I recall, the woman in the front seat was the worst off."

"I was still drifting in and out of consciousness," Thérèse Fournier explained. "But those men came into the bush to help us and they were so kind. I remember them talking to me and making me feel as comfortable as they could." Fournier, whose ribs, arm and leg were all broken, was in severe pain, all the time. And the puncture of her lung made her breathing both laboured and painful. Every SAR Tech there knew she was close to death.

"The pilot was conscious when we got there," Charlie Fleming explained, "but he had a fractured femur and two broken ankles as well as the cut on the side of his face. He had not lost a lot of blood though, because the night had been pretty cold and the blood had congealed."

"I remember first seeing those men," said Fournier. "And then I guess I must have blacked out again. When I came to they were holding me under the arms and lifting me out of the plane. They told me that when they moved me it would hurt, and it did — a lot — but they were so gentle."

"That poor woman was in bad shape," Al Williams remembered. "But she never complained because she knew we were doing our best to help her. I remember her moaning as we lifted her out. Even though she was in terrible pain, she was really brave."

As Thérèse Fournier was being freed from the wreck, a Huey helicopter carrying SAR Tech Mike Byrne had arrived overhead. In no time a litter was lowered, Fournier was strapped into it, and she was hoisted into the air.

"I can still remember the fresh air on my face on the way up," she explained. "That's really when I first knew I was not dead. Then men in the helicopter talked to me and we flew away, but I was pretty sore so the ride was not so great."

A Huey is not a large helicopter, so only one litter patient at a time could be taken. Fortunately, the hospital in New Liskeard was only a 15-minute flight away, so as soon as one crash victim arrived there, the crew returned for their next patient. By the time the helicopter was back, the men on the ground were ready to hoist.

"We had a lot of trouble getting the two women out of the back seats," said Al Williams. "Fortunately we had the crash axe. It works like a can opener in a situation like that. We also had a saw with us to cut through cable. But just getting at the folks in the back seat was a challenge. I remember working through one of the little windows there, trying to stabilize the one on the left, as two of our guys crouched inside what remained of the plane, trying to ease her out. We were afraid at the time that her pelvis had been broken, so we did all we could to lessen her pain.

"All of the people were cold, because after all, they had been there, jammed together and unable to move, for over fourteen hours. They were all somewhat hypothermic, and their veins were constricted too much to start intravenous. We were really lucky that it was such a short distance to hospital by air."

When all the victims had been taken to hospital the helicopter returned for the SAR Techs. They left the scene in good spirits, with the knowledge that four people had survived because of what had been done.

And Thérèse? She flew again, but because she never completely recovered from her injuries, she will never forget that ill-fated first flight.

CHAPTER TEN

Never Give Up

"Nobody dies on my watch!"

Coast Guardsman Don Murray was based at Cape Cod, Massachusetts when a wild nor'easter hit the New England coast late in the afternoon of April 16, 1996. The storm swept in from the cold Atlantic and roared across Nantucket Sound and Cape Cod Bay, lashing everything in its path. Sheets of wind-driven rain obscured visibility and kept almost everyone indoors. The few cars that moved, moved slowly, windshield wipers slapping in a futile attempt to get rid of the rain. When Murray's wife and three children dropped by the station to bring him dinner, their car was almost blown off the road.

As the Murrays talked, the rain pounded the hangar where they stood, and the gale blew away everything outside that was not tied down. "As we were talking the alarm went off," Don Murray said. "A fishing vessel was in trouble, ninety miles [145 kilometres] or so off the Cape."

The craft was called the *Pauline and Pearl*, a 23-metre-long steel-hulled boat that had been heading for the small city of Gloucester, a fishing port on Cape Ann, northeast of Boston. Its captain, 37-year-old Carl Chaput, had radioed for help when the heavy seas became too much.

"We've got serious water," was the understated way

Chaput described his plight when he called the Coast Guard. Towering waves had been washing across the stern of his trawler, and because it was heavy with fish anyway, the added weight made matters worse. Chaput and crew members Lennie Dow and Jeff Matthews pumped for all they were worth, but the seas got ahead of them. All three knew they would die without help.

Back at Air Station Cape Cod, Carolyn Murray and the children headed for home, while her husband and his crew fired up their Jayhawk helicopter, obtained tower clearance and flew off into the teeth of the storm. Launch time was 5:17 p.m. Thirteen minutes later the *Pauline and Pearl* reported their aft (rear) compartments were completely flooded.

"Our trip out to the boat was difficult," Don Murray said. "The weather was still terrible, the visibility horrible, and we had to fly at treetop level. Because the station is several miles inland, we went over houses. Later that night, there were complaints from residents about low flying. But we had to use the roads and houses to make our way to the water. It took something over half an hour to get to the scene.

"At that point, the [*Pauline and Pearl's*] stern was awash and the boat was still trying to make land. But with every other wave it kind of sat on its rear end. It would barely crest each wave.

"We had taken out a bilge pump for [Chaput], but we waited to see if he could make port without needing it. As time went on however, the waves got bigger and the wind stronger. There was a lot of heavy, heavy rain, and then it got dark. There was no need in even trying to get the pump down."

As the chopper crew assessed Chaput's situation they became more and more concerned. Finally Jayhawk co-pilot Mike Shirk, who had been in radio contact with the *Pauline and Pearl*, decided the time had come. He told the fishermen to get into survival suits, attach strobe lights to them, and prepare their life raft. He also called a second fishing boat that was several kilometres away to ask if it could come closer to render assistance. In the meantime pilot Andy Berghorn did his best to hold the helicopter about 20 metres above and just behind Chaput's vessel.

"[The boat] was riding up the crest of a huge wave," Don Murray said, "but when it went down the backside, it had heeled to port. We watched as the guys on the boat climbed out the pilot house windows. Because of the tilt they were actually standing on the side of the pilot house. A minute or so after that, the three of them jumped overboard." No lifeboat could be used.

Now Murray knew he would soon be in the water as well.

"I decided to do what we call direct deployment," he explained. "I hooked my harness up to the hoist cable. I then was going to take a sling, wrap it around them and bring them back up. At this stage the rescue basket was not going to be used."

As Murray made his final preparations to be lowered to the men in the water, flight mechanic Kevin Monroe checked and double-checked the swimmer's

Rescue swimmers go through extensive training to prepare them for dangerous water rescues like the one these men are practising. (U.S. Coast Guard)

gear, taking extra care to make sure the hoist cable was properly hooked. When the helicopter became momentarily stabilized Murray swung out into the night and Monroe began to ease him down. Up front the pilots fought the controls to try to remain in place.

"When I went down," Murray said, "it took Kevin a while to get me into the general vicinity of the three guys because they were all huddled together around a life ring. Finally I got down and reached out to grab one of them. But just as I touched him the wind pushed the helicopter a good distance away. When that happened, I got jerked out of the water, carried about a hundred yards [90 metres], then dropped back into the water. Fortunately I was still hooked to the cable."

The strain on Murray's body was intense, as it was on the hoist cable, because during this wild ride the cable came taut against the Jayhawk's doorframe. Try as he would, Kevin Monroe could not ease the lifeline out to keep it unobstructed. He did his best, but positioned as he was in the open door of the helicopter, he had to balance himself or he too could have been thrown into the sea. He wore a harness to prevent such an accident, but the constant and unexpected lurching was always a concern.

"As soon as I hit the water again," Murray said, "I started swimming toward the men. I finally got back to the same guy [Jeff Matthews], told him that we would be hoisting and that he would be the first to go. . . . I got the harness around him, got it real tight.

"The toughest part, though, was the helicopter was going all over the sky and the waves were high and there was slack cable lying in the water. My biggest concern at that time was to make sure the cable didn't get wrapped around anything. I knew if it was . . . and it got jerked up, it would easily tear an arm off, or even decapitate one of us.

"It took a while, but the slack was finally taken up. Then both of us got jerked out of the water, but we started a pendulum swing. As we did the swing, we hit the crest of a huge wave. Suddenly we were underwater, and my mask was torn off. I remember thinking, Holy smokes, this is not good. Jeff and I were right together and I was kicking as hard as I could to reach the surface.

"During all this time I had no communication with the chopper. Only hand signals. But Kevin finally got us up to the helicopter and he helped me get the guy inside. By the time we got him in though, and I was without my mask, I looked down and there was blood all over the floor.

"Of course, I wondered whose blood it was, and realized it was Jeff's. When we got yanked out of the water the second time, the hoist hook had hit him in the face. But I gave him a thumbs up, and he gave me a thumbs up, so I knew he was all right."

During the recovery operation several of the wind gusts were in excess of 70 knots (130 kilometres per hour), so doing another hoist recovery might be too risky. But it was not just the gale force winds, the driving rain and the instability of the helicopter that brought about Don Murray's next problem. Jeff Matthews was yelling something that Murray at first missed. He yelled again: "You've got to get Lennie. He can't swim."

"When he told me that," Murray explained, "I looked at the flight mech, and he looked at me and we decided to use the rescue basket."

This time Monroe was able to place Murray close to the two remaining men, who still clung to the life ring, despite the monstrous waves that crashed over them, often sweeping them far from the place they had been only one wave earlier.

"The rescue basket was being lowered to us," Murray went on, "and I was holding onto Lennie in case he got washed away from the ring. I also hung onto the captain because I wanted them together. The captain was okay, but I could tell that Lennie was becoming lethargic, while at other times he would be combative. Every so often he would panic and start flailing, and just go crazy. All the while I'm trying to keep him calmed down. During those times he got the attitude that he might as well relax, he was going to die anyway. He was up and down, like a roller coaster.

"Finally the basket came down and I shoved him into it. The basket has a narrow opening, so with a large [person] in a survival suit, it was kind of hard to cram him inside, especially with the thirty-foot [9 metre] seas coming at us, in the dark, and in my case with no mask, so the rain was pelting me in the eyes. I got him into the basket, though, and he gave me the thumbs up and I was yelling 'Hold on, hold on tight!' But then when they started to pull up the basket on the cable, the pendulum effect began again and the basket started swinging. Then it hit the base of a wave and Lennie fell out."

Murray was exhausted by this time, but he knew that unless he got to Lennie right away, the fisherman could drown. "You hold onto that ring!" he yelled at Captain Chaput. "I'm going to get Lennie."

Murray swam as quickly as he could the 50 metres to where Lennie had dropped into the sea. He grabbed the big sailor, then, kicking for all he was worth, hauled him back to the life ring and told him to hang onto it.

"By this time, I was screaming at him," Murray said, "and I told him that the next time I got him into that basket, he was to stay there. I told him to hang onto the thing!"

By now pilots Berghorn and Shirk were taking turns flying the helicopter because the constant pitch and roll of the machine made it almost impossible to hold in a hover. Neither pilot had any fixed reference anymore because the doomed fishing boat had drifted out of sight, so they dropped flares to try to give themselves some kind of reference point. But the flares were often of little help. One minute the chopper would be 50 metres above the waves. Two seconds later it would drop to 20, then turn on its side and be blown backward by the violent wind. It was like riding a bucking horse in a canoe. One of the pilots threw up as he fought the controls.

"Finally they were able to get the basket back to me," said Murray, "and I stuffed Lennie into it again."

As quickly as he dared, Kevin Monroe activated the hoist and the basket started to rise.

Again it swung wildly back and forth, at times on its

side, at times almost tilting upside down. Once it careened under the helicopter, slammed into the bottom and gouged a gaping hole in the metal skin. Throughout everything Lennie Daw hung on, contained his fears and was at last hauled inside to safety.

But now a new problem emerged.

When Monroe went to send the rescue basket back down the hoist would not work. He pressed the control button several times, but nothing happened. The hoist cable had become badly frayed and had birdcaged, or bound itself, around the hoist drum. He desperately tried to fix the thing, but to no avail. The hoist was broken.

"While all this was going on, the captain and I are down at the life ring," Murray continued, "and I was telling him to be sure to hold on once he was in the basket. I knew he understood because he had seen what happened to Lennie.

"And then we waited, and waited, and waited, but the basket wasn't coming down. I was getting angry, wondering what the holdup was. I had one of our little waterproof radios with me so I called Kevin. I'm yelling, 'Send the basket down! Send the basket down!' But nothing happened. It turned out that I was transmitting but I wasn't receiving. Kevin could hear me but I couldn't hear him. The radio was a piece of junk anyway, and that was the problem.

"So here we are, in the water, and I'm telling the captain, 'Carl, I'm getting mad. The basket is not coming down, and I've no clue how we're going to get out of here. But believe you me we'll get out of here, I just

Don Murray refused to give up while saving the lives of three men on the doomed Pauline and Pearl. *(U.S. Coast Guard)*

don't know how yet.' At this stage, I thought the helicopter would be dropping us a raft. We'd hang onto it and then they would send another helicopter for us.

"All the time I'm holding onto him and holding onto the life ring and trying to figure out what was going on, and getting mad. So then we started talking about family. So here we are, in these thirty-foot [9 metre] seas and pounding rain and I can't see a thing, and we're in freezing cold water, so we started to talk about family to keep our minds off just how bad things looked. Then I remember telling him: 'Hey man, nobody dies on my watch!'

"But all this time, I was still wondering how we were going to get out of there. Every once in a while I'd cover my eyes and turn around into the wind, looking for help anywhere. And then one time when I was right up on top of the crest of a wave, I saw, right on the horizon, a little tiny light. As soon as I saw it, I said, 'Okay

Carl, I know how we're going to get out of here, but it's going to take time. That's why we're talking.'"

The light belonged to the only other boat in the entire area, but it was the trawler that Mike Shirk had called, long before the helicopter rescue attempt had even begun.

The vessel was the *Maru*, owned by Mark Theroux, one of Carl Chaput's good friends. In a remarkable display of seamanship, Theroux brought his boat right to the men in the water. He hauled them on board, and ten hours later delivered them safe and sound to the port of Boston.

There were no fatalities in the entire incident, except the *Pauline and Pearl*. It sank to the bottom of the storm-tossed Atlantic.

CHAPTER ELEVEN

Dive into Death

*"From the moment I was sucked in,
I pretty well knew I was done."*

On July 16, 1991, Steve Sykes almost died at a dam on the Severn River.

Sykes was a professional diver, but his regular job was as a firefighter in Toronto. "When you work for a fire department, you have rather strange hours," he said, "so every time I had the chance I would do some diving for a company in Hamilton. The boss there, Leif Soderholm, asked me if I would go up to Muskoka and do some work on a dam for the Orillia Water, Light and Power Commission. I said, 'Sure.'"

Sykes and the two men with him, Dave Babcock and Dave Devalk, pulled their van up beside the Swift Rapids Generating Station along the Trent-Severn water system. "The people there told us they were losing hydraulic fluid in one of the lines going to a dam gate," Sykes recalled. "They wanted us to go down and try to find the leak and repair it. The idea was that I would dive into the chamber behind the gate, and then air would be pumped down the hydraulic line. The air would escape wherever the break in the line was, and I would essentially follow the bubbles to see where repairs had to be made."

The dam chamber Sykes would enter was like a large water-filled concrete vault, about 15 metres deep.

On the downriver wall of the vault, near the bottom, was a 9-tonne steel door called a headgate. By raising or lowering this door, the amount of water flowing through the dam into the power generating turbines could be regulated or stopped. But because of the break in the hydraulic line, the headgate could not be moved. It was stuck in the DOWN position.

Before entering the water Sykes went over the procedure with his crew. His backup that day was Dave Babcock. Babcock, too, was fully suited, ready to enter the vault if Sykes ran into trouble. Dave Devalk was the tender. His job was to look after the umbilical, or lifeline — a combination of safety rope, air line and communications wire taped together — which linked Sykes to the surface. The tender payed out the line to the diver as he descended, and kept in constant radio contact with him.

"It was almost mid-morning by the time I was ready to go into the water," Sykes recalled. "That's when the dam operators assured me the gate was closed and everything was in place. As soon as they had air flowing through the line I went into the chamber. I could see the bubbles and I began to follow the hydraulic line down to check where they were coming from. There was also some oil in the water. . . . As I went down the light got dimmer and dimmer, until everything was black. I had a light on my helmet, a very bright sabre light, but that was it."

Just above the headgate is a large, curved concrete abutment. Sykes swam down to the abutment, and then realized that the bubbles were coming from below

Steve Sykes spent six hours trapped under water before rescuers were finally able to free him.

it. He made his way farther down. He couldn't detect any change in pressure or any water flow at all.

Then he went deeper, almost 12 metres underwater, keeping his eye on the bubbles and feeling the wall of the abutment as he descended. At the bottom he moved in for a closer look. His light swept along the floor of the chamber to the great metal headgate, and his umbilical snaked below him.

"Ordinarily, the umbilical . . . kind of floats along with you as you descend, and the tender feeds you line as you need it," Sykes explained. "You don't want the line too slack . . . or too tight. . . . When the diver gets to wherever he's going to work, the tender allows a bit extra so the person in the water is not hampered in whatever he's doing."

At the top of the dam, tender Dave Devalk felt the tension on the line and kept paying it out into the chamber. He knew Steve Sykes had to be down about

The Swift Rapids generating station where Steve Sykes was trapped underwater. (Brian Burnie and Orillia Water, Light & Power Commission)

as far as he would be going, yet the line was still taut.

"I was right at the bottom," Sykes continued, "and I turned to check my umbilical."

What he saw horrified him. Instead of drifting or even floating upward, the line leading away from him was being drawn toward the bottom, because the headgate was not completely closed. Even worse, the umbilical was being sucked underneath!

"I yelled at Dave, 'Pull up! Pull up!' but he couldn't pull up. The water pressure right at the bottom was unbelievable. The gate was open about ten centimetres or so because — as we learned later — some rocks had become wedged under it."

While the other men struggled with the line, desperately trying to raise it, Sykes tried to help himself.

113

He moved closer to the gate, braced himself against it, and reached to pull his umbilical free. That attempt almost killed him.

The tremendous pressure of the water being forced through the tiny opening hit Sykes from behind, slamming him hard against the steel door in front of his face. His hands, both on the lifeline, were yanked forward into the outflow, momentarily trapping him. He tried to pull back and succeeded, but both his gloves were ripped off by the water and his right knee and left leg became wedged in the space below the headgate. Fortunately, neither went right under it.

"I was trapped," said Sykes. "I knew it was about fifteen metres up to the hatch opening, and that I was under about twelve or thirteen metres of water. I figured getting out of here would likely be chancy, but I still wasn't about to give up.

"I told the surface to send me down another line so I could attach it to one of the D-rings on my harness and maybe they could pull me up. When it came down I hooked it on and then they tried . . . but it didn't work. Even with my harness on, the top of my body was being pulled, but my legs were not moving. . . . I knew if they continued [pulling], they would have pulled my body apart. I told them to tie the line tight though, because it held me up. . . . The force of the water made it impossible for me to move either of my legs at all."

Back at the surface, Babcock was preparing to dive in and help his friend, but Sykes insisted that he not try it. One diver in trouble was enough, and Babcock would not be able to help if he came down anyway —

he would be in danger of being trapped too. Babcock finally agreed, and from then on he, Devalk and the other people present wracked their brains to come up with a way to free Steve Sykes.

Sykes remained kneeling against the steel door in the oily blackness. To save power he turned off his dive light. "I really thought I was going to die there," he admits. "In fact, from the moment I was sucked in, I pretty well knew I was done. I decided that if this was my time, then it was; I had no regrets. I was content with the way I had lived my life, and I had pretty much done what I wanted. I'm not a religious person, so I didn't pray. Instead, I thought of my mother and my sister and my nephew and of the good relationship I'd had with all of them.

"I didn't have a watch on, so I had no idea of time. . . . I never knew how long I'd been there, but I remember the first while being the worst. I was in unbelievable pain; and the position I was in didn't help. But then I started to lose all sensation in my legs, until eventually I could not feel anything from my waist down. . . . I realized that if I ever did get out I would probably be paralyzed, so my life was going to be changed no matter what.

"One of the first rules of either commercial diving or scuba diving is to not panic if you get in trouble. That's easier said than done, but I forced myself to remain calm. Being able to talk to the surface helped. I knew they were doing everything they could for me."

By now Ontario Provincial Police officers were on the scene, along with firefighters, ambulance attendants, local emergency measures officials and technical experts from the Orillia Power Commission. Among these were John Mattinson, an electrical engineer, and Brian Burnie, a designer draftsman whose knowledge of the workings of the generating station would prove to be invaluable, but still no one had come up with a way to free Steve Sykes.

Then someone mentioned search and rescue.

"I certainly remember the call," said Colonel Rick Hardy, who was at RCC in Trenton that day. "Luckily we had a helicopter and some guys in wetsuits training in the Bay of Quinte at the time, so they were tasked to Muskoka immediately."

"There was a lot of confusion, and quite a crowd had gathered when we landed," SAR Tech leader Mike Simpson said. "I was on Rescue 308. Steve Ackland was the other SAR Tech with me, and we knew another Lab with three more of our guys was on the way. As it turned out, we needed everybody.

"Neither Steve nor I knew anything about this kind of thing," he said, "so we needed a crash course in how the dam worked. . . . One of the engineers there told us the gauges they used showed that the gate was closed, but that debris was apparently under it. That's why the guy was trapped in the first place.

"[Sykes] was terrific, and really helpful. I had an awful lot of admiration for him because he was really brave. We were able to talk to him and monitor his condition, and what it was like down there. I'm sure he was aware of the stats for survival, though. A bystander told us there had been several incidents like this in the pre-

Mike Simpson (shown at left), along with Steve Ackland, positioned the tarps that would stop water from flowing into the chamber where Sykes was still trapped.

vious five years, but in every case the diver had died. That sure didn't raise our morale much."

On the way to the site, Simpson had been in radio contact with the Defence and Civil Institute of Environmental Medicine in Toronto, and with Toronto General Hospital. Doctors there told him that even if he was able to get Sykes free right away, he would have to remain at a three-metre depth for thirty minutes in order to decompress. If not, he could die from the bends.

"I remember looking down through the opening in the top of the dam, and all I could see was black, oily water," Simpson said. He asked the people there what they had tried so far. "As they told me, I thought each idea they'd come up with was good and I would have done the same things. Only none of them had worked. . . . I knew we would have to go down inside. By this time, our second helicopter arrived on scene."

At this point, the rescuers turned their attention to a

part of the underwater chamber called its stop log wall. Located upriver, about ten metres from the headgate side, and made of large wooden beams placed on top of one another, the wall helped control the amount of water passing through the dam. The more logs in place, the less water could move through. When the chamber where Steve Sykes was trapped was empty, the stop logs held back virtually all the water coming down. On this day, because the chamber had already been full of water, beams placed in the wall would not seal themselves, so water continued to gush between them. The rescuers had to figure out a way to close the gaps and reduce the pressure on Steve Sykes.

They had always known that sealing the cracks would help, but so far no workable way had been found to do so. But then somebody wondered aloud about finding some big tarps. The firefighters who were there located two very large canvas tarps and lots of sand bags. It was a start.

"Steve Ackland and I got into the water above the stop logs," Mike Simpson said, "and the folks there hauled the tarps up onto the dam and then lowered them down to us. The three SAR Techs who had just arrived acted as tenders while we did the diving."

Ackland and Simpson draped the tarps down over the wood wall. Then they heaped sandbags on the bottom ends, to keep them pressed tight against the logs. The pressure of the water also helped to flatten the canvas against the wall, holding it in place. Sandbags on the top of the dam kept the top edge of the tarps from being dragged down by the current.

The entire operation had taken about 45 minutes, and the water in the chamber began to drop.

"It wasn't very fast," said Mike Simpson, "but it was dropping maybe ten centimetres a minute. The dam operators told us the chamber was narrower down farther, so we figured the drop would speed up eventually. In the meantime, we were concerned for Mr. Sykes."

"I knew when I began my dive that none of the logs had been in," Steve Sykes said, "but Dave Devalk said they were put in after I got trapped. Then when I heard they had some tarps holding the water back, I felt better. I knew that if this worked, then it would be just a waiting game for the water to drop. I was glad the SAR Techs were there, and I was really impressed by everything they did that day."

"From time to time, Mr. Sykes reported feeling nauseous, and that was a constant worry," Mike Simpson continued. "We were afraid he might throw up, and that would have really complicated things. As it was, we were feeding him oxygen to reduce his chances of decompression sickness if he had to be removed quickly. . . . Despite the fact that Mr. Sykes warned us not to dive down to him, we felt we had to in order to see what we faced." Two divers put on scuba gear and went into the vault.

According to situation reports, the rescue teams were somehow under the impression that Steve Sykes's feet had actually passed under the headgate. Because of this, they made every effort to keep the gate in its present position. Otherwise, they feared, the main gate could suddenly work itself loose and crash down, severing the diver's legs. To hold the gate in position, the divers went down inside the headgate abutment and attached cables to metal rings on the top corners of the gate. These extremely dangerous dives were made by Will Bruce, with Marty Maloney acting as his safety man. Bruce had to go down alone, headfirst, in total darkness, into a space that was barely wider than his shoulders.

"They had a line on me to keep me from being sucked farther down," Will Bruce said, "but when you are upside down, the dive regulators don't work as well and I kept breathing a mixture of air and oily water. That was the hard part. I did become a bit more apprehensive the longer I was there.

"Marty Maloney was holding onto my feet and he had warned me not to go below where he couldn't reach me. But while I was inside I actually saw Steve Sykes's light go on for a few seconds. I thought of going down to him but decided not to."

Eventually the cables were connected to the rings, and then in turn hooked to a crane boom on the surface, but the crane could only lift about three tonnes and was far too small to hoist the nine-tonne gate. For this reason a complicated rope-and-pulley system was put in position for added security in case the gate moved.

Gradually the water level in the chamber dropped to chest height, so Will Bruce went in to check the trapped diver, supported by Brian Weir. Weir also took a small hydraulic jack down with him, in hope that it could be wedged under the headgate in order to raise

Will Bruce managed to secure the headgate, to keep it from crashing down and severing Steve Sykes's legs.

it, but the jack was two centimetres too high.

"As the water level dropped I started to feel a decrease in pressure," said Sykes. "I remember putting my hands down to try to free my legs, but not being able to do so. But a bit later when I did the same thing with my hose it came out a little. I pulled more of it back, and more came. Then I tried my legs again and they came. I was free!

"I backed away from the gate and told the surface I was out. Two seconds later the SAR Techs grabbed me, but by then I don't think I was really with it. It had been a long time. . . . I know they got me into a scoop stretcher which they had tied to the crane up top. Then one of them hooked himself to the stretcher and both of us were hoisted out.

"Then the pain hit me. I guess my circulation was coming back because I felt intense pain and a lot of nausea. I was given both morphine and Gravol and then wheeled across the dam to an air ambulance. The next thing I knew it was three days later and I was in Toronto in a bed at Sunnybrook Hospital."

Steve Sykes spent more than six hours in the dam chamber. He spent the first few days after his rescue in the hospital in serious condition, having developed gas gangrene in his left leg. The infection threatened to spread. He was ultimately airlifted to Fillmore Hospital in Buffalo, New York, for surgery to contain the disease. Finally his condition improved. After more operations, weeks of rehabilitation and months away from work, Sykes returned to the fire department. He is able to walk, run and swim, and despite his ordeal at the Swift Rapids dam, intends to dive again.

CHAPTER TWELVE

Into the Freezing Arctic Ocean

"There were two guys, both totally white from head to toe from the ice that was on them, frozen on their arms and head and body."

By the time the message reached the Rescue Co-ordination Centre in Halifax on November 12, 1996, Joshua Alookie was gravely ill. He lay on his bunk, in pain, unable to eat or sleep, and sometimes vomiting blood. He was in desperate need of the medications he had been taking, the ones that would keep him alive. But there were none at hand. No doctor was available.

Alookie was a crew member on the *Vesturvon*, a Danish-registered fishing trawler working near the Arctic Circle in northeastern Canada. He had had a stomach operation three years earlier and since then had to take medication every day. But now his pills had run out, he appeared to have a bleeding ulcer, and his ship was far from shore.

"We were packing parachutes that morning," said Bryan Pierce, a search and rescue technician at Greenwood, Nova Scotia. "When the RCC called, they told us that a sick fisherman needed help up north, and that a helicopter would go from Goose Bay, Labrador. We would fly top-cover in a Herc."

When a helicopter goes some distance out to sea, a

fixed-wing aircraft accompanies it — in this case, a Hercules. The Herc would reach the ship first, establish radio contact and then assess the situation. It would also be nearby in case anything went wrong. Its presence overhead generally added a sense of security for those in trouble below — as long as an aircraft was there they knew they had not been abandoned.

"We heard that the ship might make land and we would not have to go," Pierce's partner Keith Mitchell said, "but then that changed because they were still too far out. Finally we heard that the ship would sail to Frobisher Bay, but that they didn't have the charts for Frobisher." And those maps would be necessary to navigate the deep, cliff-edged, 230-kilometre-long fjord that juts into the southeastern tip of Baffin Island.

After some consideration, officials decided that the Herc with Mitchell and Pierce on board would fly to the naval base at Shearwater, Nova Scotia, pick up the maps and take them north. They could then be packed in a waterproof container and dropped by parachute to the *Vesturvon*.

As the Greenwood Herc flew north, a Griffon helicopter took off from Goose Bay. There were four on board. They would fly to northern Labrador, then on to Resolution Island, off the tip of Baffin Island. If Joshua Alookie could be brought there he could be picked up and transported to hospital. The Griffon crew was not qualified to do a direct hoist from the ship.

But hopes of a pickup at Resolution Island were soon dashed.

"We flew over it," Keith Mitchell continued, "but

we soon realized that it would not be a good place to attempt a rescue. The people on the ship would not only have to transfer the patient to land, they would have to do so in heavy seas that were pounding the shore. Then they would have to carry the poor guy up the face of a cliff to wherever the helicopter could land. It was just a treacherous area. But if we had to, Bryan and I were ready to parachute to the island and look after Mr. Alookie until the helicopter could pick him up."

As the Hercules flew above the ship, freezing rain and snow showers swept the surface of the sea. Lights from the *Vesturvon* became a tiny beacon in the desolation, but even that beacon often disappeared in the swirling snow. The navigator, Carol Elliott, kept in regular contact with the ship, but it seemed that each time she inquired, Mr. Alookie's condition had worsened.

"As we circled the ship we had an urgent message from the Griffon," Bryan Pierce explained. "They told us that they had encountered severe weather, strong head winds, and that their visibility was limited. They wanted to land somewhere to wait out the storm, but they couldn't see where to put down. They asked us if we could drop some flares so they could find a place to land."

"We hated to leave the ship," Keith Mitchell said, "but when the helicopter radioed that they needed help, we went to them because we felt their immediate situation was worse. We knew that the sick man was at least safe on the boat, but that the helicopter had to have help right away."

Captain Mario Defoy, who was at the controls of the Herc, flew immediately to the area above the Griffon. His crew dropped several flares from about a thousand metres, providing enough light for the Griffon to land. The Herc went back to the fishing boat. By now Joshua Alookie's condition was extremely serious.

"Each message from the ship was worse than the one before," Bryan Pierce said. "They told us the sick man was severely dehydrated, incoherent and close to death. They said that if he didn't get help right away, he would die."

Because the helicopter was down, any hope of using it to collect Alookie in time was gone. As the snow streaked past the windows of the Herc, those on board quickly discussed the limited options available. If the ship sailed to Iqaluit, the nearest place where a doctor was available, Alookie wouldn't last. If he was taken to Resolution Island, both he and those taking him might be smashed to death on the rocky shore — and without a helicopter available there would be no way to pick him up anyway. The third option — having SAR Techs Mitchell and Pierce parachute into the ocean — was extremely dangerous. The storm continued in the blackness outside, the seas were churning and the terrible cold the jumpers would encounter once they hit the water could render them immobile. In other words, if the two SAR Techs jumped they would be facing death themselves.

Then another problem arose: the plane was almost out of fuel.

"So we knew we had to make a decision right

away," Keith Mitchell explained, "and even though neither Bryan nor I wanted to jump into the Arctic Ocean, we felt that that was about the only option left."

"From then on it was a mad scramble in the back of the plane," said Pierce. "Keith and I had to strip off our Arctic clothing and get into wetsuits. Then our gear had to be put into plastic bags and sealed. The charts for the ship had to be waterproofed. We had to have our medical equipment ready. We taped flashlights to our legs, got our little inflatable one-person life rafts ready, put glow sticks on our helmets so we could see each other in the dark, and double checked our parachutes.

"The crew was helping us, of course, and we worked like crazy. I sweat a lot, and at one point I had to take off my boots and pour the water out. Unfortunately, when the ramp door was lowered, the floor of the plane became an instant skating rink."

Keith Mitchell continued the story: "On one of our passes over the boat we had seen a Zodiac [an inflatable boat with motors] on deck, so we had Carol Elliott call down and tell them to get it in the water and be ready to pick us up. Apparently the guy she talked to on the radio thought we were absolutely crazy to be making the jump in the first place, but he said the Zodiac would be ready. As it turned out, a couple of seconds before I jumped I saw the thing — still on the deck. That didn't do a lot for my confidence at the time.

"We were wearing black wetsuits and orange vests, which are themselves a floating device. We had swim fins of course, and we taped these to our boots so we would not lose them either on the way down or when we hit the water. When the back door was opened, the pilot flew over the ship, and our night light, which is a drift indicator, was tossed out. We could see it flashing and flashing, but then it went out. On the next run the second light also went out, but at least it gave us some indication of how strong the winds were."

Then the plane came around for one last pass over the drop zone. The two SAR Techs stood at the edge of the open ramp at the rear. Outside was the absolute blackness of the night, broken only by the white streaks of snow. The roaring engines and the shriek of the wind made speech impossible.

The two men leaped from the plane at an altitude of 600 metres, one kilometre back from the ship. As soon as they were in the air the full force of the Arctic gale hit them, driving both down and forward at about 90 kilometres an hour. There was little time to do anything but ride the wind, and it almost became Pierce's undoing.

"One of the things you have to do in a water jump is to release some of your buckles before you hit," he explained. "But because of the bulkiness of my wetsuit I had trouble releasing my chest strap." Pierce tugged at it for about ten seconds, but because he was going so fast, he overshot where he wanted to come down. "The first thing I knew I was a kilometre or so away from the ship, so I turned into the wind to reduce my speed as much as I could. But then I was down. The entire jump took less than a minute."

When a jumper lands on the water his parachute usually settles on the surface. But this time the chute stayed inflated. The winds caught the canopy, jerked

Pierce forward, then began to drag him, face down, in a wild ride through the towering waves. The shock of the cold ocean, the power of the wind and the darkness on every side were overwhelming. Pierce fought desperately to gain control, all the while being dragged farther and farther into danger. He knew that if the Zodiac from the ship could not find him, he would be lost forever.

"Finally I realized what was happening to me," he said, "and I managed to grab the release mechanism and pulled it to cut the chute away. But when I did my reserve parachute came out. It didn't fill with air because I was in the water, but the problem was all the canopy lines were tangled around me. This made it very hard to get my gear sorted out because the waves were falling down on top of me as well. Eventually I got my medical kit and the rest of the stuff I needed, but I had to let my parachute go and swim away from it.

"I could see the lights of the ship in the distance, but I knew that even if I had abandoned all my gear I would not have been able to swim that far. When I was on top of a wave, I could see, but when I was down in a trough I could not see anything. That went on for a while, and it was [then] that I really started to notice the water temperature." Pierce knew he had better figure out how to get out of the freezing water.

"Because I didn't know how long I was going to be there, and because of the cold, I decided to inflate my life raft. It inflated as it should, so then I threw my gear into it, held onto it for a while, then climbed up into it. I remember actually laughing . . . because it was such an effort to just get out of the water."

Pierce's partner, Keith Mitchell, landed more or less where he intended, on the lee side of the ship, where there would be some protection from the wind. But the protective element actually turned into another hazard. The sudden drop in wind strength caused his parachute to collapse, then inflate, then finally come down on top of him in the water. In such a situation a jumper can suffocate, because it is impossible to breathe through the canopy fabric.

"I also had all the shroud lines tangled around me," Mitchell said, "and it took a while to free myself. Then I put my hands over my head and tried swimming to get out. That didn't work at first and I thought I might have to cut a hole in the canopy and climb up through it.

"But as I pulled my knife out I remembered that the chute was worth ten thousand dollars, and I had better not wreck it. So I put my knife back in the sheath and just kept moving . . . until I at last got a corner of the canopy off me.

"I decided not to inflate my life raft because I was afraid if I got into it I might be flipped over by the wind. So I stayed low in the water, but this was not so great either. I was holding my gear, the waves were crashing over my head and splashing in my face, and sometimes I couldn't breathe. When that happened I would turn around and grab a mouthful of air, but the salt water would splash in my face and I would swallow some of it with the air. After a while I swallowed so much I got sick. I'd throw up, then a wave would roll over me and my own vomit would crash into my face.

Because of the cold, it was like ice pellets hitting me.

"Then it just became a matter of holding on. I was getting pretty cold, but I was worried about Bryan. I knew I was closer to the ship than he was, and I wanted to get to him, to make sure he was okay. I remembered that on the jump he had hit the water [first], then I hit, but I hadn't seen him since.

"One minute I would be at the crest of a wave where I could see the ship, then that wave would pass and I was down in the dark — bobbing up and down and shivering. I was actually getting farther and farther away from the ship, and I knew that unless somebody from the boat got to us pretty soon we would be out of sight and they would never find us.

"But just as I was thinking that I saw the Zodiac zipping through the water. Then it started slowing down because it was obvious they saw me. There were two guys, both totally white from head to toe from the ice that was on them, frozen on their arms and head and body. And ice was encrusted on the pontoons on the sides."

As the Zodiac bobbed up and down, surrounded by floating ice, Keith Mitchell was hauled aboard. When he stood, momentarily, the lukewarm water in his porous neoprene wetsuit ran out and he was left exposed to the deadly lash of the Arctic wind.

"I started shivering uncontrollably," he said, "but I was really worried about Bryan. The guys on the Zodiac didn't know where he was, and for a time neither did I. But after we sailed around, looking, I noticed his light. When I knew they had seen him, I got down below the sides of the pontoon, out of the wind. They brought him on board, and then the two of us curled up together, shivering. Man, was it cold!"

The sailors in the Zodiac quickly manoeuvred their craft alongside the *Vesturvon*, and assisted the freezing SAR Techs up a rope ladder onto the ship's deck. From there they were taken to the bridge to warm up, and minutes later began to work on the sick man. First they gave him Gravol and morphine intravenously, then co-ordinated by radio-phone with medical specialists in both Iqaluit and Halifax to determine which drugs would best improve his weakened condition. Fortunately the treatment worked and Alookie began to respond. Just 15 hours later, by the time the trawler had sailed up Frobisher Bay to Iqaluit, Joshua Alookie was actually coherent, and even able to walk.

But the men who saved him could not rest. They only had time for a quick interview with the media before flying off to search for four other rescuers who were missing in Labrador. After a fast meal they were taken directly out to the Herc that was waiting for them, its engines running.

Twenty-two months later Bryan Pierce and Keith Mitchell were awarded Canada's highest award for bravery, the Cross of Valour, for their heroism in saving the life of Joshua Alookie.

Keith Mitchell (left) and Bryan Pierce wear the Cross of Valour, awarded for their courage in saving Joshua Alookie's life. A Canadian Coast Guard officer described the jump as one of the most heroic rescues ever undertaken.

CHAPTER THIRTEEN

Plunge into Darkness

Then they began stumbling through the snow, ice and total blackness toward what remained of Boxtop 22 . . .

The northernmost settlement on earth is at a place called Alert, a Canadian military post on Ellesmere Island, 800 kilometres from the North Pole. In 1991, Alert was the scene of a terrible plane crash that would take five young lives and scar several others forever. It involved a Hercules transport plane, one of three flying in supplies from Thule, Greenland.

These flights were made each year, but were often interrupted by the quickly changing polar weather. "Boxtop" was their code name.

Late in the afternoon on October 30, Hercules #322, or Boxtop 22, was on its way to Alert from Thule, flown by John Couch, a 32-year-old Gulf War veteran and a seasoned Herc pilot. On board were 4 crew members and 13 passengers. The plane was also carrying 15,000 litres of diesel fuel in a tank in the cargo compartment. It was a routine flight in good weather and good visibility, despite the darkness. The temperature that day was a seasonable minus 22 degrees Celsius.

As he neared Alert, the pilot talked to the control tower. He could already see the runway lights in the distance, so he chose a visual, rather than instrument-controlled, approach. But because of the total darkness — Alert is so far north that it is in total 24-hour darkness from mid-October until March — he was flying what is sometimes called a "black hole" approach, one where the pilot cannot use the horizon as a visual reference. Flight simulator tests have shown that most pilots "crash" when attempting such an approach. In Couch's case, the decision to make a visual approach was tragic.

The plane began its descent above the frozen surface of the Lincoln Sea, immediately beside the airstrip at Alert — or so the crew thought. They were actually still about 15 kilometres out, over land that was dotted with hills. By the time anybody realized this, the plane was so low that it hit one of the hills.

The port wing touched the ground, and with an incredible screech of tearing metal the plane lurched to the left and broke apart. The wreckage ended up scattered in a crude semicircle across a shallow depression surrounded by outcrops of rock.

The wings of the plane came to rest in one place, the tail in another, the cockpit in a third. All four of the plane's propellers were ripped off. After the sound of tearing metal stopped, there was an explosion of aviation gasoline and the roar of fire. Flames lit the daytime darkness as the plane was reduced to charred rubber, melted metal and indescribable pieces of wreckage. Bodies, living and dead, were tossed into the snow and drenched in reeking diesel oil as the internal fuel tank tore loose and broke open. The flight was over.

Amazingly, there were survivors. They were scattered: a couple near the cockpit, some behind the wings, two more lying off to one side. They were stunned and disbelieving.

A Hercules similar to the ill-fated Boxtop 22 that crashed at Alert.

But help was already on the way. When the flight did not land on schedule, the control tower tried to locate it, visually and by radar, unsuccessfully. Because another supply plane, Boxtop 21, was also en route to Alert, it was redirected by the tower to the last known position of the missing plane.

At 4:50, twenty minutes after the first plane was scheduled to touch down, Boxtop 21 saw fires on the ground.

Ten minutes later search and rescue was called.

❖ ❖ ❖

Warrant Officer Arnie Macauley had just sat down to dinner in Greenwood, Nova Scotia. "I was still eating when my brother Marvin phoned," he recalled. "All he said was, 'There's a Herc down up in Alert. Get everybody in. It looks as if it's a MAJAID.'" Marv Macauley was a Hercules pilot with search and rescue. MAJAID was a military code term for Major Air Disaster: the crash of an aircraft with more than ten people on board. Arnie immediately left for the Base.

"One of the guys at work called me," SAR Tech Ron O'Reilly remembered. "He said there was a Herc down . . . and that they needed us all in. At first I thought he was pulling my leg, but he said, 'No. It's for real.' I got dressed and broke a land speed record getting in."

The same thing was happening with SAR Techs across three time zones.

"It was just after three-thirty in Edmonton," said Darby Darbyson at the search and rescue unit there. "When the buzzer went off we jumped up and went right over to the Ops Room in another hangar to find out what was going on. The guy I talked to said: 'Get on your plane right now. There's a Herc down at Alert.'"

Every search and rescue squadron was on the alert, but help was still far away. There were no helicopters or SAR Techs nearby, so men and equipment were launched from Greenwood, from Gander, Newfoundland, from Trenton, Ontario, and Edmonton, Alberta, and later even Iceland and Alaska. First off the ground were the teams from Trenton and Gander, coming by helicopter. Soon fully loaded Hercules rescue planes from Edmonton and Greenwood were on their way too.

The distances were daunting. The crash site was about 5000 kilometres from Toronto. In fact, it was closer to Moscow. It would be a seven-and-a-half-hour flight from Greenwood, a little less from Edmonton.

"The lack of information was a problem at first," explained Greenwood Team Leader Arnie Macauley. "We knew there'd been a crash, and that someone had seen flares near it, but at one point we were told that the plane might be down on the pack ice, and that there could be open water, so that meant we had to bring equipment for a water rescue, along with Arctic gear. We also loaded toboggans, water, personal survival kits, flares, drop lights, parachutes — everything we thought we might need. In all we had fourteen SAR Techs on board. I left four behind on standby.

"We knew as we went along that Edmonton was on the way as well, and that their ETA [Estimated Time of Arrival] would be an hour before us. We figured then that by the time we got there Fred Ritchie would have a bunch of guys on the ground, so we began concentrating on camp gear and so on."

Darby Darbyson was on the Hercules from Edmonton. "About an hour out of Edmonton we were told that a storm was approaching Alert. Then by the time we got there, the weather was really bad. It was storming, the clouds were low and you couldn't see a thing. We flew over where we thought the crash was, several times, but nothing was visible on the ground. We threw out flares but they didn't help at all. They just reflected off the clouds. You'd throw a flare, turn away for a second, and when you looked back you couldn't see the flare. It was very windy, snowing heavily, and when we had the ramp down at the back, really cold."

Finally the plane landed at Alert. They needed to wait for a break in the storm before trying again to spot

Arnie Macauley,
Greenwood team leader.

the crash site. Fred Ritchie decided to split his team in two. He would lead a group on the ground, moving by tracked vehicle. A similar attempt earlier, by base personnel, had been halted by a huge gorge southeast of Alert, so Ritchie planned another route. The rest of the crew waited to take off again in the Hercules, as soon as the weather cleared enough.

Meantime, the Greenwood team was on its way.

"We were expecting to hear notice of a crash location," said Arnie Macauley, "status of survivors and whatever, but the next thing we heard was that the Edmonton plane had landed at Alert. . . . We did know that a storm had been moving in, that things might not be good, but we never thought it would be so bad we couldn't get down. . . . We couldn't see a thing anywhere. And the area under us was so featureless anyway. We saw snow on snow and nothing more, even with the ramp open and everybody looking out, in the back and up front. Then by this time, we were getting short on gas, but there was no room at Alert, so we flew to Thule. We knew that the guys from Edmonton were going to try again."

"We went back up, but it wasn't much better than before," Darby Darbyson explained, "and we kept flying and flying. But then we got some great news from the ground! We began to pick up a transmission from the crash site. Before their batteries down there gave out we learned that there were fourteen survivors originally. But after a while [we] knew that somebody had died. To save their power, a system was worked out so that the radio just had to be clicked. One click, ten alive. Two clicks, twelve alive, and so on.

"The big boost to us was knowing that there were people alive down there, and we wanted to get in to help them. We tried everything we could. There were guys looking out back at the ramp, out the spotter's windows, up in the cockpit. At one point we dropped drift lights, but they were gone immediately, so we knew the winds were really bad. For a second we thought we'd seen the site, but then we lost it and never saw it again. And we were never sure if it was the crash or just black rocks. By this time we were completely burned out, so we went back to Alert."

On the ground at the crash site, the survivors were unaware of the bustle of activity in the air. They knew they were in desperate straits. They concentrated on coping with injuries and keeping warm. They radioed for help, counted heads, and prayed. They knew they were close to Alert, so they expected rescue soon.

But the rescuers did not come. As the hours dragged on in the darkness, and the cold and pain continued, their spirits fell. Mere survival became the priority. Two of the survivors, Bob Thomson and Susan Hillier, remained separated from the group, too badly injured to be moved. In time they would be buried in snow. The people who were mobile made their way to a section of the back of the plane which had stayed mostly in one piece. They lay down, huddled together against the bitter cold in their metal cave.

"When we landed at Thule, we figured it was all over for us," Arnie Macauley continued. "We expected the Edmonton crew to get in, either in the air or on the ground. . . . But old Fred was going to go out again and they needed us to drop flares to guide them. So away we went again. That was when our ordeal really started. We piled into the Herc and did a real white-knuckle takeoff out of Thule. There was a cross wind, almost to the limit of the C-130. My brother had to pull off the line there to keep it on the runway, so we did basically a kind of three-engine takeoff. Finally he got the thing airborne."

"When we got back over Alert," said Ron O'Reilly, "it was clearing near the ocean, but inland there was still a blizzard. Then we could see the lights of the tracked vehicles [from the search party on the ground] and had a fair idea about where they should go, so we started dropping flares to guide them. Every so often we could fly back over the crash site to see if we could see anything. We never could though."

Macauley continued: "We established radio contact with Fred and his guys and we guided them as best we could. But they couldn't see fifteen feet [4.5 metres] on the ground. Their compasses weren't working and they had ended up on sea ice earlier. We didn't want that to happen again, but guiding them was a lot of work. We did seven or eight hours of that and it was hard on everyone. We'd fly along, then the guys on the ground would get stuck, and a couple of times they went crashing over cliffs even when they had walkers out in front. When they got bogged down we would go over to the crash site, home in on a beacon that was operating down there and see if we could see anything. We still couldn't, but we did get the ground party over the Sheridan River and onto flat ground and thought they would be our best bet at getting to the crash. They were perhaps three to five kilometres away.

"All this time Fred was navigating by the altimeter on his watch because nothing else worked. He knew the crash was at about twelve hundred feet [360 metres], so he knew when the ground started to rise and he got to twelve hundred feet he might be there.

"Then, one time, when we went back to where we thought the crash was, we were in the right place at the right time. We'd just dropped a flare, there was a slight clearing in the storm, and my God, we saw the tail! . . . There was such a sense of relief for all of us on board. But then we came around again and never saw anything for another hour, and in all, we had been involved in this thing for about thirty hours at that

Months after the crash, when daylight returned, the wreck of Boxtop 22 could be seen. (Bruce Fraser)

point. Now though, everyone knew something was going to happen."

"We didn't have enough headsets to go around," said Ron O'Reilly "and because you can't hear anything in the back of a Herc, some of us were really pretty much in the dark as to what was going on. We were just sitting there waiting for the order to jump.

"During all this time we were dropping flares, and the ramp at the back was open. Because the guys there were getting cold we switched every so often. I went back for a while along about then and had a chance to plug into an intercom outlet and hear what was going on. I remember Gerry Dominie was in the window, I was on the ramp, and Arnie was looking out as well, and Gerry saw something like a tail. Arnie saw something also and they asked me. I said I sort of saw a black object, but I wasn't sure whether it was a tail or not. Then we had another look and it was definitely a tail. That's when we got ready to go."

"We pretty much ignored the ground party at that point," Arnie Macauley explained. "They were still some distance away, and it didn't look as though they would get there soon, so I started briefing my guys on what we would be doing.

"We would go in in three different sticks [groups of jumpers]. The first stick was going to be six people with strictly medical gear, and they were going to go in and set up a triage to treat the life-threatening injuries first, and so on. The second team was going to be responsible for the perimeter, to get the camp set up and retrieve the gear. The third team would stay on the airplane, drop all the gear and they would jump afterward. The plan was okay, but then it fell apart.

"Just as the first stick was about to go the storm was moving out of there and we couldn't see a thing. But when we did see the tail again our own contrails [exhaust] caused so much vapour in the air that everything would just fog over and cloud everything up again. At one point we got off course when we thought we were at about a thousand feet [300 metres], and we saw one of our flares bouncing along the ground right under us.

"There was a lot going on in the front end, and the guys were working really hard to get the aircraft back over the spot again. Tensions were very high, lots of comments on the intercom, and everybody really frustrated.

"And of course, we all wanted to get on with it because people were dying down there, and the whole world was watching. This has been going on for well over a day now and nobody's in at the crash yet. Here we are, flying around, the ramp's open, it's forty below out there, and a hundred knot [185 kilometres per hour] wind is whipping around in the back of the plane. From the wheel wells aft, everything is frozen solid, we're scraping the ice off the windows, and the guys up front are cooking because we had to have as much heat as we could.

"Finally we dropped another flare, saw the crash and decided this was it. Ordinarily we never would have jumped in those conditions under two thousand feet [600 metres], or in those winds. But we decided that the first six jumpers would go out, then a flare would be dropped right after them in the hope that it would come down somewhere near them and give them a bit of an idea of where the ground was. The flare burns for two minutes at two million candle power. Finally we went out at a thousand feet [300 metres]."

Arnie Macauley was first out of the plane. "It was eerie coming down. The flare was an orange glow, and we could see it, but we could sense nothing else at all, other than an unbelievable blast of ice crystals. It took about forty-five seconds to come down, and at the last instant I saw the tail of the crashed Herc right beside me and I thought I was going to hit it. We had drifted almost two miles [3 kilometres] coming down. We were really moving.

"One of our other jumpers, Bruce Best, was number two, and he came out of our plane behind me, but he hit the ground first. I saw him bounce, then the wind took his canopy and he was off like a shot, so then he collapsed his canopy right away. I remember hitting hard and the back of my head snapped into the ground with a thud. Then I felt myself getting picked up as the canopy started to take me away. It took a bit of struggle to control it."

They had arrived.

Macauley gathered his crew around him. They secured all the equipment and radioed up to the plane to report their success. They set off, stumbling through the snow, ice and total blackness, toward the wreckage of Boxtop 22.

"At this point I don't think any of the survivors

knew we were there," Macauley continued. "We swept through the wreckage, yelling as we went, but there was no response that we heard. Two guys went on the left, two more in the centre, and Ben House and I were on the right. The guys on the left found a couple of bodies along the way. As we went we kept yelling to keep from getting lost. The storm was so wild that you could lose a guy fifteen feet [4.5 metres] away — and we had lights on our helmets. We had radios, but we didn't dare use them because they were pretty much useless. Three or four transmissions in that cold would kill them.

"I came up to the tail section and yelled in there, and for the first time got an answer. The tail was wide open except for a life raft that was bent up on one side a little, but the whole place was packed with snow. The snow was swirling around, and even with my headlamp, I couldn't see much. I got down on my knees because there was a lot of sharp pieces of metal and junk, and for the first time I realized the floor was covered with people. You couldn't move without stepping on somebody. One guy, whose name I found out later was Mario Ellefsen, was at the feet of everyone, and I kind of fell into him and he screamed. Later we found out he had a broken pelvis."

Macauley also found a survivor, Paul West, who told him there were two others outside in the snow. "He thought the last contact with them was seven or eight hours before," said Macauley, "but he wasn't even sure of that. . . . Two guys stayed at the tail to do what they could and Ben and I went out with Bruce

Ron O'Reilly survived the harrowing jump into the crash site at Alert.

Best and Derek Curtis to try to find the ones outside. As we went we were kicking at stuff in the snow, and I remember pulling up about five parkas, because you couldn't tell if it was a person or not. Unfortunately the survivors never found any of this stuff when they could have used it. This tells us something about the trauma of the crash on the folks who were in it."

The four SAR Techs did a sweep up one entire side of the wreck and found no one, but then two of them found Thomson and Hillier covered in about a metre of snow. The two were alive, but in bad shape.

"I remember being back at the tail of the plane again when the second stick — five guys — came down," Macauley recalled. "As they got closer the parachutes were silhouetted in the night from the flare they had just dropped. In a way, it was a beautiful sight, with all the ghostly-looking orange crystals in the air. When the guys landed they all grabbed their chutes pretty quickly and then came over to us."

"As we came down I remember the yellow glow of the flare, and the ice crystals in the sky had this sparkling look," said Ron O'Reilly. "We could see the ground then, but the snow was still blowing. I could see the tracked vehicles and it looked as though they would arrive at the crash site as soon as us, but it was still a while. As we got closer to the ground, we could see dark objects and white objects, so I decided to stay away from the dark objects because they were likely rocks.

"Once you were on the ground it was a blizzard, and the only things you could see were the strobe lights of the other guys. I remember cutting out of my risers so I wouldn't be dragged away in the wind, and then grabbing my chute right away. . . . I thought that if I couldn't find the crash site, at least this would be some shelter. Even though we were all together, you could be in a total whiteout in two minutes. I think we all thought the chutes might be needed if we got lost."

"The last three guys didn't get to jump," Arnie Macauley explained. "They did one more pass and threw the equipment out, but because they were on bingo fuel [low in gas] they had to get to Alert and land right away. Unfortunately, the equipment they dropped was all lost — two hundred and eighty thousand dollars worth of it. I know because I had to do the complete inventory later. They dropped it in the right place,

MAJAID toboggans like this one were dropped at the crash site by parachute, but most blew away before the rescuers could use them.

but the winds were so strong they took it all. We did see a couple of big bundles go by, but they went as fast as a horse could gallop. We had no way of stopping it.

"The situation in the tail of the plane was really bad. I knew those people had lain down for the last time. None of them could walk; not even Paul West. They were frozen to the metal floor, with a combination of sweat at first, and condensation, then urine. If we had not been able to get there when we did . . . We had a terrible time trying to break them free from the metal. Some we just took out by cutting their clothes. They were in such a mess, but we had to worry about the hypothermia as well as the various injuries. By this time we had found another body outside, and pilot

John Couch as well, who was lying beside Mario. We removed the body at that point. Meanwhile, some of the other guys were working with Sue Hillier and Bob Thomson outside."

Back at Alert, the crew of the Edmonton plane had been fed and given places to sleep. But nobody got much rest at all during those hours.

"I was in a bed next to a phone," said Darby Darbyson. "I don't think I'd even closed my eyes when it started ringing. I grabbed it of course, and somebody told me the Greenwood guys were on the ground and needed help.

"We were out of there, and three of the Greenwood SAR Techs who didn't jump earlier, Keith McKellar, Gerry Dominie and Marc Lessard, climbed into our Herc and we took off. But the weather was still terrible over the crash.

"We jumped from about eight hundred feet [240 metres] and we couldn't see a thing. Jim Brown went out first, then myself, and Shawn MacDiarmid came out behind me. I saw Jim hit the ground, or what I thought was the ground, and I saw Shawn hit. I hit, and even though I had a good landing, I blew into Jim, chute and all.

"When we got organized we didn't know where we were, and about that time we started to wonder what we'd gotten ourselves into. We later found out we were about two miles [3 kilometres] from the crash site. It was in the middle of a blizzard, dark, and you could see nothing at all. So we got on the radio to the aircraft and asked them to have the guys at the crash fire off a flare to give us direction, because not one of us knew where to go, nor had any one of us seen the crash as we came down.

"So we stood in a circle, looking out, like a bunch of muskox, but we saw the flare off in the distance. In order to keep from getting lost we decided to leapfrog toward it. Two guys would walk a short way, then two more would walk to them and go a bit beyond. Then two more would come up and so on. We kept doing that, lining each pair up, and yelling back and forth all the time to keep from getting lost. Every so often we'd ask for another flare to keep us in the right direction. We always had to talk to the plane and they talked to Arnie, because there was a hill between us and the crash site.

"We came to a ridge line . . . which the aircraft had hit. Then we walked down a slope, into a bowl and came to the site. There was a mass of metal everywhere, but apart from the tail, it was hard to tell that it had been an airplane. The whole scene was so strange. There was no light other than the headlamps of the SAR Techs who were there, and occasionally the light from a flare up above that our Herc was dropping to guide the ground party.

"The first guy I saw was Ron O'Reilly, and I remember asking him if there was anyone still alive, the place looked so bad."

The SAR Techs at the scene have jumbled, blurred memories of the next few hours. Some men remember some things in vivid detail, but much has been forgotten. They were all desperately tired, but kept going long after people are supposed to keep going. So many things had gone wrong — the continuing storm, the lost ground parties who nearly perished themselves, the last group of SAR Techs who might have died on the trackless tundra, the equipment that was lost before it could be used. But now, the problems were ending.

"They began to end when a couple of my guys, Jean Tremblay and Rob Walker, found a toboggan jammed into some boulders somewhere," recalls Arnie Macauley. "They brought it in, took a tent out of it and were beginning to set it up, just as the last group of jumpers came over the hill. The tent was a six-man SAR tent . . . and there was also a lantern. They set the tent up, got the light in there, and finally we had some place to put Hillier and Thomson, to get them some protection. At about this time Fred Ritchie and the two tracked vehicles from Alert arrived.

"We had been at the site for two hours and forty-five minutes when the ground party pulled over the horizon. I know when I saw old Fred I was never so happy to see anyone in my life. They had more tents, cases of pop, chocolate bars. Coleman stoves, all the things Base Alert had given them. We set up more tents, got some people melting snow for water, started to warm up our IVs because they were frozen solid, and began moving people inside."

"The tail section was an unbelievable scene," Darby Darbyson said with a shudder. "The parachutes were over the opening when I went there, and then inside there was another chute over the people. It was hard to distinguish where they were, and you didn't want to step on people. I'll never forget the moaning. It was like walking into a meat freezer. Then everyone would be absolutely silent.

"I went to one guy, Mario Ellefsen as I later found out, and started to feel around to see what I could do for him. He was in severe pain, and his clothes were frozen to the ramp. He said his back hurt, and this was because his pelvis was broken. We had to try to figure out how to get him loose without hurting him any more than necessary. Finally we shot him up with morphine, cut his clothes off, put him into a sleeping bag, and carried him to one of the tracked vehicles. Ben House started an IV in him and we got him into MAST pants [anti-shock trousers] to stabilize his fractures. At least now his broken bones weren't grinding on one another. He was conscious all the time."

"Once we got one or two out of the tail, we had a bit more room in there," Arnie Macauley said. "But before we were able to give them morphine, we had to be able to monitor them because they were in such a sorry state. Morphine is a respiratory depressant, so if you give it to somebody who is too weak, you could stop them from breathing and kill them. That's why we had to be so careful. And of course everyone in the tail had been drenched in the diesel fuel, and that, along with the sweat and condensation and urine made it so bad for them all. That was before you even considered their broken bones and burns."

"I remember working on one guy who was in bad shape," Ron O'Reilly recalled. "He told me he had to pee in the worst way. I told him to go ahead, that we were going to cut his clothes off anyway. I guess that okay wasn't the best, because right afterwards he relaxed and was right out of it. Then he didn't respond anymore. I knew he likely had a fractured skull, but I didn't know whether it, or hypothermia, was the bigger problem. Anyway, we got his clothes off, I did a rectal temperature check to see what his core temperature was, and we got him into a sleeping bag and started an IV.

"And we were really getting burned out by this time. I'm not sure of the number of hours, but it seemed like a couple of days without sleep. When you were outside, you were in the cold, but after working in the warmth of the tracked vehicles for a while, the adrenaline rush began wearing off and you felt so sleepy. I remember trying to get an IV into one person, but I was falling asleep and I couldn't see the vein. I got someone else to do the stick for me and I got out for a few minutes in the cold to wake up."

While his men worked to save lives, Arnie Macauley was on the radio. He knew that a helicopter was coming in to lift out survivors, and he wanted to prepare the personnel at Alert for the casualties they would soon receive. Finally the helicopter, a Twin Huey, arrived.

"We knew the Twin Huey could only take four stretchers at a time, so we picked the most serious for the first flight out," Macauley explains. "I think the next was two stretcher cases and three survivors who were ambulatory, and so on. Anyway, we got all the injured out in three lifts. Six SAR Techs had gone with them, and then the Huey broke down.

"Fortunately it didn't cause a great problem because we had been told two American Pavehawks [helicopters] from Elmendorf [Alaska] had been taken to Thule and they would be in to help us bring everyone else out. Well, the Pavehawks arrived okay, but because of their size, when they landed their rotor wash smashed down all our tents and broke them up. We were lucky we didn't need them anymore.

"But at the very end there was another problem. I wanted to send everyone who was left out on the last flight — including the bodies of the five victims. The American pilots didn't want to take the bodies, so I talked to somebody higher up in Alert. Alert talked to the pilots and the bodies were taken."

So the long, cold, exhausting ordeal was over for the SAR Techs and their support personnel. Their actions in that frozen hell were incredible. Without them, not a single person on Boxtop 22 would have come out alive. But every man involved in the operation, in one way or another, said simply, "It was my job."

The wreckage of Boxtop 22 stands as a monument to the lives that were lost when the Hercules crashed. (Bruce Fraser)

aft: the rear of a plane or ship

altimeter: a device on an aircraft showing its altitude

Billy Pugh: a mesh/scoop rescue device

bingo fuel: the point where an aircraft is almost out of fuel

Buffalo: type of aircraft

C-130: Hercules transport plane

chopper: helicopter

CPR: cardio-pulmonary resuscitation

contrail: condensation trail in the exhaust of an aircraft

ELT: emergency locator transmitter

ETA: estimated time of arrival

flight deck: the cockpit of a plane

flight mech: flight mechanic

H-21: a type of helicopter, sometimes called a "flying banana"

H-60: a Jayhawk helicopter

helo: helicopter

Herc: Hercules transport plane

homer: electronic homing device

Huey, Twin Huey: types of helicopters

hypothermia: abnormally low body temperature

IFR: instrument flight rules

IV: intravenous

knot: 1 nautical mile per hour

Lab: Labrador helicopter

lee: the side sheltered from the wind

MAJAID: major air disaster

MAST pants: anti-shock trousers

GLOSSARY

Mayday: an international distress call

medevac: the transportation of sick or wounded patients by air to hospital

Night Sun: high-powered light on search and rescue helicopters

OPP: Ontario Provincial Police

ops: operations

port: the left side of a ship or aircraft when facing forward

prop wash: wind originating from the propellors

RCC: rescue co-ordination centre

RCMP: Royal Canadian Mounted Police

rescue shock: a condition in which a victim ceases to fight to stay alive because rescuers have arrived

SAR: search and rescue

SAR Tech: search and rescue technician

stick: a group of parachute jumpers released rapidly from an aircraft

Stokes litter: a device, similar to a stretcher, for lifting a patient

triage: the process of determining the order in which ill patients will be treated

VFR: visual flight rules

Zodiac: an inflatable rubber dinghy, especially one with an outboard motor

Photo Credits

Paul Beattie: 51

Brian Burnie and Orillia Water, Light & Power
  Commission: 113

Canadian Forces: 17, 25, 30, 69, 70

Lisa DeGroot: 61

Bruce Fraser: 143, 157

The Houston Chronicle: 42

Tyler Jennings: 87

Mike Johnston: 49

John Melady: 12, 18, 39, 72, 73, 93, 112, 117, 121, 133,
  136, 139, 148, 150

The Montreal Gazette: 74, 77

The Sitka Sentinel: 89

U.S. Coast Guard: 5, 9, 83, 102, 108

The Vancouver Sun: 20